T0031897

ANCIENT BONES

— Foreword by David R. Begun —

Madelaine Böhme
Rüdiger Braun · Florian Breier

TRANSLATED BY **JANE BILLINGHURST**

ANCIENT
BONES

Unearthing the Astonishing New Story of How We Became Human

GREYSTONE BOOKS
Vancouver/Berkeley

First published in English in 2020 by Greystone Books
First paperback edition 2022
Originally published in Germany as *Wie wir Menschen wurden*
Copyright © 2019 by Wilhelm Heyne Verlag, Munich, in the Random
House GmbH publishing group
English translation copyright © 2020 by Jane Billinghurst
Foreword copyright © 2020 by David R. Begun

22 23 24 25 26   5 4 3 2 1

All rights reserved. No part of this book may be reproduced, stored in
a retrieval system or transmitted, in any form or by any means, without the
prior written consent of the publisher or a license from The Canadian
Copyright Licensing Agency (Access Copyright). For a copyright license,
visit www.accesscopyright.ca or call toll free to 1-800-893-5777.

Greystone Books Ltd.
greystonebooks.com

Cataloguing data available from Library and Archives Canada
ISBN: 978-1-77840-031-5 (pbk)
ISBN: 978-1-77164-751-9 (cloth)
ISBN: 978-1-77164-752-6 (epub)

Copyediting by Lesley Cameron
Proofreading by Jennifer Stewart
Cover and text design by Nayeli Jimenez
Front cover artwork by Hauptmann & Kompanie Werbeagentur, Zurich,
Switzerland
Front cover illustrations from The Jane Goodall Institute, USA (footprint
of chimpanzee) and Shutterstock/Kristo Robert (human footprint)
Interior illustrations by Nadine Gibler Informationsdesign
Artist's renditions by Velizar Simeonovski
Photographs as credited on page 322

Printed and bound in Canada on FSC® certified paper at Friesens. The FSC®
label means that materials used for the product have been responsibly sourced.

Greystone Books thanks the Canada Council for the Arts, the British Columbia
Arts Council, the Province of British Columbia through the Book Publishing
Tax Credit, and the Government of Canada for our publishing activities.

Canada

Greystone Books gratefully acknowledges the xʷməθkʷəy̓əm (Musqueam),
Sḵwx̱wú7mesh (Squamish), and səlílwətaʔɬ (Tsleil-Waututh) peoples on
whose land our Vancouver head office is located.

CONTENTS

FOREWORD

I AM HONORED TO call Madelaine Böhme a friend. I have never met anyone quite like her, and I have met just about everyone working on the fossil evidence of ape and human evolution over the past forty years. Her knowledge of the fossil record and geology are encyclopedic and her enthusiasm for research is infectious.

Madelaine and I had been following each other's work for some time before we finally met in Istanbul at a conference in 2013. Shortly after that, she invited me to work with her on an intriguing fossil, El Graeco, which she described with the unique combination of historical documentation, geological and paleontological expertise, and personal passion that typifies this book. The invitation came at a particularly exciting time for me, as I had been promoting the idea that Europe was not the backwater of ape evolution, contrary to what was widely claimed by many researchers, mostly those working in Africa.

Under Madelaine's direction, our work on El Graeco has brought the hypothesis that Europe had a central role in great ape evolution back to the attention of our skeptical colleagues. It has also revealed her dogged determination to extract as much data as possible from the most unlikely of sources. I had studied this fossil in the mid-1990s, when it was on loan to the Natural History Museum in London, England. I concluded that it was unique, but I did not know what to make of it at the time, given its somewhat poor state of preservation. By using high-resolution X-rays (micro computed tomography) to look inside the fossil, Madelaine and her student Jochen Fuss were able to identify characteristics that distinguished El Graeco from other fossils from Europe and linked it with hominins—that is, humans and human ancestors that evolved after our lineage branched off from that of the chimpanzee. Madelaine's genius is encapsulated by her realization that she could date the El Graeco site from a specimen hidden away in a long-forgotten archive using the orientation of crystals inside the bone. Who would have thought of that?

The narrative that Madelaine Böhme unveils in this book is as compelling as it is controversial. She brings together a wealth of information from history, geology, paleontology, and archeology to call into question some of the most treasured ideas in paleoanthropology, dogma in fact, about ape and human origins. She is refreshingly free of the biases that have led many to ignore data that contradict prevailing hypotheses. Her book sheds much-needed light on alternative interpretations that are supported by new discoveries but met with predictable skepticism. All the details of Madelaine's narrative may subsequently be challenged as new data and new analyses are published, but she has opened a new frontier of research and is asking new questions, long repressed by the paleoanthropological establishment. Many of the ideas expressed in this book will one day find their way into textbooks on human evolution and inspire readers and students alike to ask themselves, "Why didn't I think of that?"

Madelaine describes some of the difficulties that paleo-anthropologists working in Eurasia confront when their evidence points to interpretations that differ from the widely accepted view that African apes and humans first appeared and evolved exclusively in Africa. Though the "Africanist perspective" can be traced back to Darwin, the great evolutionary biologist was more open-minded than many of our colleagues today, suggesting, as Madelaine notes, that the ape from Europe known as *Dryopithecus* may be a relative of African apes and humans. Darwin was marvelously uncertain, as any scientist should be in the absence of direct evidence, of what he called the "birthplace and antiquity of man." We should all be so open-minded. More importantly, Darwin recognized the significance of what we call in today's parlance "land mammal dispersal events" or, as Darwin put it, "migration on the largest scale." An important theme in this book is about dispersals, the movement of animals, including apes, back and forth between Eurasia and Africa. We know this happened often. In fact, the animals that typify the African savannah today mostly arrived from Greece and Turkey. *Graecopithecus* or its kin were probably among them. Animals do not carry passports and do not worry about borders.

Confirmation bias (seeing only data that support one's own world view and ignoring contradictory data) is not confined to paleoanthropology. It is ubiquitous in all fields of knowledge and tends to pit researchers against one another. Readers should guard against falling into this trap. This is not

XII | ANCIENT BONES

a competition between Africa and Eurasia for prominence in the narrative of ape and human evolution. Indeed, in the period leading up to World War II, many leading researchers, all men of European ancestry, strongly defended the idea that Europe was somehow a more suitable place for the development of advances in brain size, intelligence, and culture than other parts of the world. Few if any researchers would support any of this ideology today, but we are nonetheless still confronted with the problem of confirmation bias. Fortunately, Madelaine's outlook, a recurrent theme of this book, is resolutely positive. I am certain that readers will be left with the same emotion as Darwin, who famously started the final sentence in his magnum opus with the words: "There is grandeur in this view of life."

I look forward to working with Madelaine, our colleague Nikolai Spassov, and their students and colleagues for many years to come, adding to the increasingly impressive amount of data revealing the central role of the ancient occupants of Europe and Asia, as well as Africa, in the story of ape and human evolution.

DAVID R. BEGUN
Toronto, Canada

INTRODUCTION

W HEN I WAS twelve, my parents gave me a book for young people whose title roughly translates as *How Humans Rose Above the Animal Realm*.[1] I was completely and utterly fascinated. The book aroused my curiosity, and I could not get the subject out of my mind—in fact, it enthralls me to this day. I found myself disagreeing with the authors' thesis. It worked its way into my subconscious, and it drove me to ask lots of questions: How can we rise above something we are part of? Isn't that an egocentric way of looking at the world? And what about "rising above"? Hasn't *Homo sapiens*, the "rational being," been a disaster for this planet? What makes us different from other animals? What makes us unique? What made our enormously dynamic evolution toward civilization possible? Which evolutionary adaptations paved the way for the arrival of modern humans, and what do these adaptations mean to us today? And last but not least, what are the facts and what is merely speculation?

Those questions have informed my work as a scientist from the very beginning of my career. They taught me that scientists must always question what they think they know. For me, that meant questioning the commonly accepted idea that apes evolved into humans in Africa and nowhere else. I had my first doubts about this theory in the summer of 2009, just as I was stepping into a new academic position, and numerous new finds and investigations over the past decade have done nothing but add to them.

The science of human evolution is currently developing more quickly than almost any other field of scientific research. Hardly a month goes by without reports of spectacular discoveries or new research results that raise questions about prevailing beliefs. An ever-growing number of innovative techniques in the biological sciences are now being used to investigate the geological, biological, and cultural developments that led to the appearance of humankind, and much knowledge that was accepted as authoritative just a few years ago is now being questioned. It is a fascinating time to be researching the evolution of humans, because many academic theories paleoanthropologists have embraced for decades are in the process of being overturned. In these times of change, I was keen to make the new knowledge accessible to a broad general readership, rather than simply my academic colleagues. I wanted to make the story of scientific discovery as exciting for lay readers as my research has been for me over the past few years. Hence this book.

New discoveries were made throughout the time I was writing this book, and keeping up with them and adding them to my text proved to be quite challenging. One discovery in particular—a discovery my research team and I made of a previously unknown species of great ape that we unearthed, piece by piece, near Kaufbeuren in the Allgäu—caused great excitement and provided amazing new insights into the evolutionary process of becoming human. The full significance of this find is not yet known, but we do know that it is one of the most significant paleontological discoveries ever made in Germany. As all this was unfolding, I brought two science journalists, Rüdiger Braun and Florian Breier, on board, and between the three of us we were able to complete this wide-ranging and challenging book project in a relatively short time.

Ancient Bones invites the reader on a forensic investigation into the origins of humankind. The goal of this book is not only to impart knowledge but also to make readers curious about the connections between evolution, climate, and the environment, connections that we may not fully understand until some time in the future. It offers new insights that contribute to our understanding of what it means to be human. It is also meant to be entertaining and easy to read. The description of discovering, losing track of, and rediscovering *Graecopithecus*, the oldest known candidate for the very first direct human ancestor, reads like a detective story. Today, I'm thrilled that I refused to stop searching, otherwise El Graeco, as we called him, would likely have remained buried forever in the sands of time.

The finds of *Graecopithecus* and the great ape from the Allgäu made headlines across the world. This book not only tells the story of how the finds came about and how they are classified in the scientific order of things—and what they have to do with German rock star Udo Lindenberg—it also gives an overview of both the shining and the shadowy moments in paleoanthropology and ends with a summary of where the research stands today.

The story of human evolution told in these pages includes much of the evolution of the great apes. It goes back more than 20 million years and includes a multifaceted picture of our ancestors from Africa, Asia, and Europe—from the beginning of great ape evolution to the development of prehumans (early hominins) and early humans to the present. There is a special focus on changes in the climate and the environment as critical drivers of human evolution. European savannahs and African deserts played as important a role in evolution as ice ages or the drying out of the Mediterranean Sea did.

This book investigates which pivotal evolutionary steps were necessary for humans to appear on this planet. It begins with the adaptations great apes made to a challenging environment, sheds light on the beginnings of upright gait, explains why early human evolution cannot have happened exclusively in Africa, and describes a world in which our species shared this planet with other species of humans.

Through all this, it becomes clear what makes humans human and how the context of our evolution explains the features and characteristics we have today: our brains, our hands and feet, our metabolism, our language, our wanderlust, and our fascination with fire. Millions of years of evolution have contributed to making us what we are today. And the task of unbiased science will always be to investigate that development, because although we have raised ourselves above the animal realm, we are still a part of it—with the gifts of insight and the ability to ask what makes us the way we are.

Part 1

EL GRAECO AND THE SPLIT BETWEEN CHIMPANZEES AND HUMANS

QUESTIONING THE ORIGINS OF HUMANS

The Detective Work Begins

I N 2009 I embarked on a scientific adventure that, in hindsight, unfolded with all the twists and turns of a mystery novel. I was about to take on a professorship. My title was quite a mouthful, but it accurately captured my area of expertise—terrestrial paleoclimatology—which means I research what the climate used to be like on land. I was going to be part of a human evolution project co-managed by the Senckenberg Society for Nature Research and the University of Tübingen. In the middle of all the upheaval that comes with taking on a new position, I got a phone call from Nikolai Spassov, the director of the National Museum of Natural History in Sofia, Bulgaria.

He and I had been friends on a professional level for many years. In 1988, when I was still a young student, I had the opportunity to participate in digs in Bulgaria with him. We were investigating sites where remains of pre-ice age vertebrates had been found. It was a formative experience for me. I found it incredibly exciting to hold the remains of creatures that were part of an ecosystem that no longer existed. Every new detail we uncovered added to our picture of the lost world and helped bring it to life. From the beginning, Spassov was supportive as I dove in.

Spassov is one of the best mammal experts I know, a walking encyclopedia of amazing anatomical facts about animals that are alive today as well as animals that died out long ago. He taught me many things, including how to recognize that the bone I had just dug up was from the upper foreleg of a saber-toothed cat or what features revealed that the numerous deer bones we had just unearthed belonged to at least three different species. He probably wondered why a twenty-one-year-old geology student had such a passion for anatomical details, especially when she was on the dig team to study the geological features of the site. But he patiently answered all my questions, and I took advantage of that. Even back then, what I really wanted to study were extinct animals and plants in the places where they had lived.

Bulgarian Inspirations

Now, over twenty years later, Spassov was on the telephone, beside himself with excitement as he told me he had finally found what he had been searching for in Bulgaria for the past ten years: the fossilized remains of a great ape—a hominid, as experts call the family to which the gorillas, orangutans, chimpanzees, and modern humans all belong.

Spassov had dug up an upper premolar that showed typical hominid features and was probably 7 million years old. That amazed me, because according to research done by many of my colleagues, great apes had died out in Europe long before then. That had been the accepted school of thought for decades, and recent discoveries from Spain and Greece seemed to confirm its credibility. Spassov's discovery, it occurred to me, completely contradicted it. What made things even more interesting was that he had made his find close to Azmaka near Chirpan in central Bulgaria, a region where no one would have expected him to find anything. The southwestern part of Bulgaria is the region known among experts for its wealth of evidence of extinct mammals.

The chances that Spassov had actually found the remains of a great ape in this area seemed as likely as his winning the jackpot in a lottery. But as I well knew, he was very good at what he did. Therefore, I agreed without any hesitation to join him on a dig that fall. Our primary goal would be to examine the geology and estimate the age of the site where the tooth was found.

For ten days that fall, Spassov and I worked intensively in the sandpit at Azmaka, along with four of my students, a small French team, and some Bulgarian colleagues. We established a geological timeline, surveyed the sediments and the sedimentary layers, and drilled rock cores in the exposed ground to get data about changes in the Earth's magnetic field that would help us date the upper premolar Spassov had found. We also found other fossils, including the almost-complete skull of an elephant. Georgi Markov was the expert on fossil proboscideans on the dig, and he recognized it right away as belonging to the genus *Anancus*, one of the first true elephants. A hominid tooth and an *Anancus* skull—up until then, such combinations had been found only in 6.5-million-year-old sites in Africa. Other species of mammals found at Azmaka also indicated that the Bulgarian site was something special. The mood of the team became increasingly excited and focused. Finally, we were able to confirm Spassov's estimated date.

IN THE LOWLANDS of Thrace in central Bulgaria, temperatures can reach 95 degrees Fahrenheit (35 degrees Centigrade) even in September. The warm evenings are sometimes the most pleasant time of day. We made the most of the balmy temperatures and met regularly in an outdoor restaurant that served authentic Balkan food. There were lamb kebabs and stewed lamb heads; traditional tomato, cucumber, pepper, and onion salad; and rakia, the local fruit brandy. We relaxed after long days in the field, and we talked.

On one of these warm evenings in Azmaka, I told Spassov about a 1949 paper by Bruno von Freyberg.[2] In 1944, the German geologist had found the lower jawbone of a great ape in Pyrgos near Athens. Its unusual features had made it difficult to classify, but von Freyberg estimated it to be somewhat younger than finds made at the famous paleontological site of Pikermi, which was relatively nearby. Many researchers had dated the Pikermi site as being about 8.5 million to 7 million years old. Scientists at the time thought von Freyberg's estimate was utter nonsense, because it was completely at odds with the generally accepted thinking that great apes had disappeared from Europe long before then. Therefore, in their opinion, there could not have been any highly developed great apes in Europe a couple of million years later. No one bothered to verify the age of von Freyberg's find.

It hit both of us at the same time. The Bulgarian premolar and the lower jawbone from Greece could have come from the same time period. Could there really be a European great ape that dated back 7 million years? That would open a new chapter in the story of the early stages of human evolution and take us into uncharted territory. I felt I was on to something. We were on the cusp of changes of sensational proportions. What better subject to investigate as part of the scope of my new duties at the University of Tübingen than this very question?

What was important now was to reevaluate the jawbone and establish exact dates for the sites at Azmaka, Pyrgos, and Pikermi. The only problem was that no one had any idea where the lower jawbone and the other fossils from Pyrgos were. And there were rumors that the site at Pyrgos had been built over and was no longer accessible. Without the fossils, and without knowing their relationship to the rocks, there was no way to undertake the scientific analysis that was needed. But I wasn't going to give up that easily. The lower jawbone had to exist somewhere, or so I hoped. After all, it had survived the chaos of World War II.

And so began a trail of discovery that would lead me back to the very beginnings of paleontology in the nineteenth century, a German army geologist in Athens during World War II, and an almost-forgotten safe.

- 2 -

THE GREEK
ADVENTURE

The First Fossil Apes From Pikermi

———

S PRING 1838. A common soldier appears before representatives of the Bavarian zoological collection in Munich and offers to sell the renowned zoologist Johann Andreas Wagner fossils from Greece. The soldier believes the sparkling crystals they contain are valuable diamonds. Even though he knows the crystals are not diamonds but common calcite, Wagner realizes immediately that the man has indeed stumbled upon treasure. There in the soldier's modest satchel, in among all the fragments of bone and horses' teeth, lies something much more precious: the upper jawbone of a fossil primate.[3]

Wagner was famous for his explorations of the "primeval world," as the past geological epochs of the world were then called. He had already studied many fossils. But there was a gap in the scientific knowledge that he and his colleagues were anxious to fill. The fossilized remains of lions, hyenas, elephants, and rhinoceroses had been found in many places in Europe and Asia, and this led the experts to conclude that these animals had once been much more widely distributed. And yet, until now, there had been virtually no ape or monkey fossils. How could it be that these species existed together in modern Africa but apes and monkeys were absent from the fossil sites? With the find from Greece, Wagner now held in his hand an important piece that had been missing from the primeval puzzle. After careful examination, he documented the find in 1839. He named it *Mesopithecus pentelicus* ("middle monkey from Mount Pentelicon") and described it as a link between langurs (Old World monkeys) and gibbons (lesser apes).[4]

But how exactly had the fossils come into the soldier's possession? That story is as fascinating as the fossils' journey to the Bavarian State Collection of Zoology in Munich. In 1836, the British historian George Finlay was combing the area at the foot of Mount Pentelicon, northeast of Athens, in search of sites from antiquity, when he came across some puzzling bones. He collected a few fossils and showed them to Anton Lindermayer, a German doctor he had befriended, who immediately recognized them as the fossilized bones of mammals.

Finlay and Lindermayer belonged to a group of Western Romantics who ardently admired ancient Hellas and called themselves Philhellenes. Their fascination with the country's past had drawn both men to Greece. Supporters of this intellectual movement included writers and philosophers such as Lord Byron, Victor Hugo, Johann Wolfgang von Goethe, Friedrich Schiller, and Alexander von Humboldt. Philhellenes sided with the Greeks in the nineteenth-century Greek War of Independence against the Ottoman Empire.

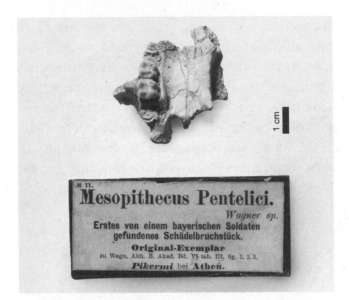

The upper jawbone found by a Bavarian soldier in Pikermi and now in the collection of the Bavarian State Collection for Palaeontology and Geology in Munich. It was used by Johann Andreas Wagner in 1839 to document a species he named *Mesopithecus pentelicus*.

An almost complete skull of *Mesopithecus pentelicus*

Illegal Greek Souvenirs

In 1827, after a civil war that lasted for many years, the guerilla fighters and the Greeks, with the help of the major powers France, Great Britain, and Russia, finally managed to defeat the Turks and bring independence to Greece. But conditions in the young republic remained unsettled, and in 1832, the major powers decided to turn Greece into a monarchy. This was the only circumstance under which they were prepared to extend the debt-ridden country the credit it so desperately needed. They suggested to the Greek national assembly that it choose a European prince as the country's king. After two other candidates had politely declined, the assembly finally settled on sixteen-year-old Prince Otto of Bavaria, the second son of Bavaria's King Ludwig I. That was a less-than-ideal solution, as Prince Otto was still a minor and therefore not yet considered legally competent. Still, on February 6, 1833, he arrived in Nafplio, which was then the capital of Greece, aboard the British frigate *Madagascar*, as the country's first king. On board were also 3,582 Bavarian soldiers and a great number of officials, including a military doctor named Anton Lindermayer. And that was how a German doctor and a British historian crossed paths in the newly made country of Greece and eventually paired up to dig up fossils.

After Finlay's initial finds, further exploratory expeditions to Mount Pentelicon revealed that a stream bed near the locality of Pikermi was particularly rich in bones. Finlay and Lindermayer hired some of the Bavarian soldiers who had arrived in Greece with King Otto to help manage the digs. One of them returned to Munich the following year with illegal "souvenirs" in his luggage. And so, the beginning of one of the most important paleontological finds to date is marked by a theft. Or to look at it another way: the Pikermi site would never have risen to international fame had that grave robber not wanted to secretly supplement his wages.

After Finlay and Lindermayer made their finds and Wagner described them, a veritable gold-rush fever gripped Pentelicon, although in this case, the prize was bones. Explorers and scientists dug in the stream bed at Pikermi, and academies and museums sent expeditions to the area. And that is why, if you visit the most important natural history museums in Europe today, you can marvel at *Mesopithecus* fossils, as well as impressive specimens of fossil giraffes, antelopes, rhinoceroses, hyenas, and saber-tooth cats. Together, the Pikermi fauna represent a living community much like the wildlife of the African savannah today—an assemblage of animals that was first discovered as an ancient ecosystem on European soil in Greece.[5]

Pikermi marked the beginning of vertebrate paleontology as an independent branch of scientific study. Many early, groundbreaking books on the history of mammals were based not only on Darwin's theory of evolution, which was published in 1859, but also on these fossil finds in Greece.[6] The fauna at Pikermi were physical proof of the idea that not only animals, but also landscapes and climates, were constantly changing. However, for the next hundred years, *Mesopithecus* remained the only species of primate ever discovered in this lost savannah world.

Detail of bone fragments from Pikermi. Embedded in the fine-grained red sedimentary rock is a chaotic assemblage of bones, primarily from horses, antelopes, and giraffes.

IN THE QUEEN'S GARDEN

Bruno von Freyberg's Discovery

WITH THE OUTBREAK of World War I, the digs at the bone-Eldorado near Athens ground to a halt. All grew quiet at the Pikermi site until a further fortuitous event rekindled interest in Greece's fossil legacy.

In 1941, a geology professor from Erlangen named Bruno von Freyberg was conscripted "for the duration of the conflict" as a geologist for the army. Army geologists were not soldiers but civil servants who worked for the army. Their responsibilities included finding suitable sites for building and securing military installations, and checking to see if the rock there could be used as building material, how drinking water could be procured, and whether there were any valuable raw materials underground.

In 1943, von Freyberg was transferred to Athens in occupied Greece. He was to document the geology north of the city, check out coal deposits, and supervise the construction of bunkers for anti-aircraft guns. Von Freyberg and his co-workers were under significant time pressure, because the situation between the German occupying forces and the local population was extremely tense. There were an increasing number of attacks and acts of sabotage by the local resistance, and Waffen-SS massacres of civilians intensified the resistance efforts.

Von Freyberg and his assistants became aware of a property with a tower and an unusual past. The grounds were known as Pyrgos Vasilissis Amalias—Queen Amalie's tower. Amalie Marie Friedorike, duchess of Oldenburg, had married King Otto I of Greece in 1836, thus becoming the first queen of Greece. Her passions were nature, farming, and gardening. To introduce modern agricultural methods to Greece, she founded Heptalophos—the Seven Hills—a 615-acre (250-hectare) estate outside the gates of Athens that was, indeed, spread over seven hills. The name was not chosen by chance; it had a symbolic meaning. Many Philhellenes, including Amalie, dreamed that one day Constantinople (now Istanbul) would once again be the capital of the Greek world and the seat of Orthodox Christianity. And Constantinople—like the original city of Rome—had been built on seven hills. Amalie had a small palace built at Heptalophos, and it was the little tower of this palace that had brought the estate to von Freyberg's attention. The geologist discovered that the southernmost hill was the ideal place on which to build a bunker for the German anti-aircraft guns.

Found in the Chaos of War, Then Forgotten

In June 1944, workers started to build the bunker. The excavation revealed something amazing. The workers salvaged a fossil from the red sedimentary rock. Von Freyberg knew right away what it was: the complete lower jawbone of a primate. More skeletal remains were unearthed, and von Freyberg was finally convinced that he had discovered an "important site of the Pikermi type," that is to say, a site with the same defining features as the one at Pikermi.

The war made it impossible to carry out a careful and thorough dig. The best von Freyberg could do was to document the geological features at the site and elsewhere on the estate. He also asked the lead sappers and workers to save any remains that were unearthed during the excavation work. Whenever he could, he dug bones out of the excavated material himself, which cannot have been easy given that he had lost his right arm in World War I. He clearly felt this was an important find.

Although von Freyberg was a geologist who also had some training in paleontology, he knew the fossils needed to be examined more closely by a seasoned expert. Therefore, he immediately sent the finds to Wilhelm Otto Dietrich, the foremost authority on fossil mammals in Germany at that time, in Berlin. Dietrich wrote back confirming von Freyberg's suspicions that the fossils belonged to the Pikermi fauna type, and he identified eleven species, including two species of giraffe and five species of gazelle and antelope. He also found one species each of horse, rhinoceros, and elephant, along with the "middle monkey," *Mesopithecus pentelicus*, the same species the German soldier had come across in 1838 in Greece.

Dietrich assigned the lower jawbone uncovered by von Frey-berg's workers to that species. This assessment later turned out to be wrong on a number of counts.

In September 1944, the German army pulled out of Greece and von Freyberg left Athens. The valuable finds from the queen's garden stayed with Dietrich at the Natural History Museum in Berlin. They were heavily damaged on February 3, 1945, when the east wing of the museum took a direct hit from a bomb and was completely destroyed. Von Freyberg survived the war and, after being cleared by panels tasked with ridding Germany of Nazi ideology and working for a while as a porter in a textile factory, he returned to the University of Erlangen. In 1950, he was reinstated as a full professor and returned to his position as head of the depart-ment of geology and paleontology. "The bones arrived in Erlangen after the war as shattered remains," he wrote in his memoirs.[7] The collection included "the now severely dam-aged" lower jawbone belonging to the primate. All the teeth on the left side and some on the right were broken off.

Von Freyberg had published his geological data about the site at Pyrgos, together with Dietrich's list of the animal spe-cies he had identified, back in 1949, but amid the ruins of wartorn Europe, hardly anyone had noticed the article, and there was no further work at the site.

That did not change until 1969, when the Frankfurt paleo-anthropologist Gustav Heinrich Ralph von Koenigswald visited von Freyberg in Erlangen. At the time, von Koenigs-wald was one of the most famous experts in his field. He had discovered important great ape fossils in Asia, as well as the early human species *Homo erectus*. When he saw the lower jawbone from Pyrgos, he realized right away, despite the damage the fossil had suffered during the war, that Dietrich had made a mistake. The thick layer of enamel and the wear on the teeth indicated that this was not a "middle monkey" after all, but a previously unknown extinct great ape. Von Koenigswald renamed it *Graecopithecus freybergi* after the site where it was found and in honor of the man who found it. Freely translated, the name means "Freyberg's Greek ape." But other scientists did not show much interest, and this time, not only was the find forgotten, but it disappeared altogether.

- 4 -

IN SEARCH OF FORGOTTEN TREASURE

A Journey Into the Catacombs Beneath the Nazi Party Rally Grounds in Nuremberg

T HE DISCOVERY OF a hominid premolar in Bulgaria was a huge scientific achievement, but just one small piece of the mosaic. I could not shake the feeling that Spassov's find and the lower jawbone of *Graecopithecus* probably belonged to the same species. But to prove that beyond a doubt, the fossils needed to be examined using modern methods. More than forty years had passed since von Koenigswald had assessed the lower jawbone. What had happened to *Graecopithecus*'s remains? Was there any chance at all of finding them?

I began my search where von Freyberg had worked until his retirement: the University of Erlangen. But none of the scientists who worked or who had worked in the department of paleontology had ever heard anything about *Graecopithecus* or the site at Pyrgos. The geology department, where von Freyberg had taught until 1962, also initially yielded no leads. I persevered for two years, and finally, on November 20, 2014, I got a message that turned things around. Someone suggested I contact Siegbert Schüffler, who had once been in charge of the geological collection at the university. Schüffler had also known von Freyberg personally, but he had been retired for the past twenty years. As it turned out, this time I had struck pay dirt.

Schüffler told me over the phone that the collection, including von Freyberg's finds, had been handed over to the Nuremberg Natural History Society years ago, with one exception—the *Graecopithecus* jawbone. Von Freyberg had personally told Schüffler that the lower jawbone was the "most valuable piece in the collection" and had asked him to look after it well. And so, back in the 1980s, Schüffler had entrusted the fossil to the secretary of the chair of the geology department and asked her to keep it in the safe.

I rang up right away to find out what had happened to the jawbone. The new secretary remembered the "monkey tooth" in the safe. It was still there in its old place in the office. When I was finally sent a photograph of the fossil, I could not believe my eyes. What a disappointment! The fossil in the photograph was the jawbone of a horse. Had the person who took the photograph got the fossils muddled up? However, I refused to give up and asked for another photograph. When it arrived, I could stop holding my breath. There was no doubt in my mind: *Graecopithecus freybergi* had resurfaced.

On December 6, 2014, I stood in the office of the secretary to the chair of the department of geology at the

University of Erlangen and watched as an old-fashioned gray safe was opened. There, inside an old plastic container from the 1980s, lay the lower jawbone the world had forgotten, the one I had been trying to track down for the past two years. Respectfully, I took the fossil out and examined it from all sides. The jawbone was smaller than I had expected. It looked slender and fragile, but the most important thing was that it was still intact enough to be helpful. The war had certainly damaged it, as von Freyberg had said, but I did not see that as a big problem. Unlocking information from ancient bone fragments was all in a day's work for me. I could hardly wait to begin the preliminary analysis.

The Remains of a Lost World in the Cold, Clammy Depths of the Catacombs

I was thrilled to have found this fossil, but there was a catch. I knew that an exact evolutionary classification of *Graecopithecus* would be possible only if I could also get my hands on the other fossils from Pyrgos. You can certainly describe an isolated find accurately and, depending on its condition, date it somewhat reliably; however, in paleontology, it is extremely important to give a fossil an "absolute date" (that is, as precise a date as possible in terms of years) and also to classify it as part of the complete extinct animal world of that geological epoch using other fossils from the site.

Von Freyberg knew that, too, and that is why he collected as many fossils as he could from Pyrgos. I therefore decided to contact the Nuremberg Natural History Society, which was supposedly looking after more pieces of this scientific treasure. This volunteer association is one of the oldest of its kind in Germany. To store part of their large collection, the officers had turned to one of the largest and most controversial pieces of real estate in Germany: the horseshoe-shaped congress hall on the Nazi party rally grounds in Nuremberg.

ON A COLD, wet day in December 2014, I met two of the group's volunteers in front of the imposing granite façade of the building the Nazi party had built for staging its monumental political events. The edifice occupies an area larger than twelve American football fields and is 128 feet (almost 40 meters) tall.

The cellars under the congress hall look like halls themselves, and they are so huge you could drive a truck through them. The walls are lined with brick, and the whole place feels eerie. The items stored in these cold, clammy conditions are here because there is no room for them anywhere else. In addition to archeological finds, there is equipment and miscellaneous furnishings.

We penetrated deep into these catacombs, walking under numerous archways, making many turns along the way—I would never be able to find my way back alone—until we were standing in front of a row of meticulously labeled brown wooden cabinets. We hit upon what I was looking for in the very first one: number A01. One of the volunteers opened the cabinet and pulled out drawer D03-1. I could hardly believe it. There it was, hidden beneath a thick layer of dust: von Freyberg's collection from Pyrgos. Yellow labels nibbled by mice were inscribed with the discoverer's handwriting: "Fauna from the Pikermi stage, Pyrgos Vasilissis, north of Athens."

In among the finds were scraps of shredded military topographical maps and a Rodina cigarette package—"Rodina" means "home" in Bulgarian. Inside the package, nestled on a bed of cotton fabric was the toothless lower jawbone of a gazelle. The condition of the material was shocking: crumbly, broken bones encrusted with sediment. A few larger pieces of rock showed no discernible traces of fossils under the gray dust and looked as though they had been collected at random. The war and postwar years had taken their toll, I thought, and then—that's going to be an awful lot of work!

- 5 -

MAGNETOMETERS
AND
MICROTOMOGRAPHY

Ancient Bones in a High-Tech Lab

I F TRUTH BE told, at the time of my visit to the University of Erlangen at the end of 2014, even the name *Graecopithecus freybergi* was being debated. In the 1990s, researchers from the United States had asked the International Commission on Zoological Nomenclature to strike the name from its records. This committee is responsible for the recognition and standardization of species names. The critics' main objection was that they doubted the claim that *Graecopithecus* was the same age as the Pikermi fauna.[8]

Their argument went as follows: Classification depended on supporting finds, which in this case had been lost, and the site in Pyrgos was no longer accessible for further digs. Therefore, a single fossil that had been damaged in the war would not provide enough evidence for recognition of a new species. If all went well, Nikolai Spassov and I would now be able to change all that. I literally held all the individual pieces in my hand, and "all" I had to do now was analyze them and connect the Pyrgos finds, the premolar from Bulgaria, and the famous fossils from Pikermi using modern scientific methods.

Our investigations began with the teeth of *Graecopithecus*, who was soon given the nickname El Graeco (the Greek) by my team to keep things simple. Teeth are of great interest to paleontologists because they are covered with enamel, the hardest organic substance we know of. That is why teeth buried in sediment often survive for millions of years in particularly good condition, and why they are often the best-preserved finds at digs. Using fossil teeth, researchers can also draw far-reaching conclusions about the ancestry of extinct animals, what they ate, and, therefore, what their environment was like.

The teeth from fossil primates are particularly informative, because they permit a clear distinction not only between great apes and other apes and monkeys, but also between great apes and the direct ancestors of humans (see the diagram on page 30). The roots of the canines of extinct great apes and those alive today are long and wide, especially in males, which use their canines to intimidate other males in fights for dominance. The roots of their molars are also longer, and each molar possesses multiple roots, normally three or four. Moreover, these roots diverge—that is to say, they spread outward. Roots that are spread out anchor the tooth firmly in the jaw in much the same way a toggle bolt anchors a screw. On the human evolutionary branch, in contrast, molar roots converge, their tips curving inward. Not only that, but the root tips of the two premolars fuse into a single power-ful root. In many of us, you can tell that the pencil-shaped roots have fused together because there is more than one root canal. The molars of great apes and humans are not the same shape, because they have different demands made on them when the animals are chewing. Different diets mean great apes and hominins (Hominini) have developed differently shaped teeth. (Hominin or Hominini is the term paleoanthro-pologists use for our species and for all our extinct ancestors who lived later than the common ancestor of humans and chimpanzees.)

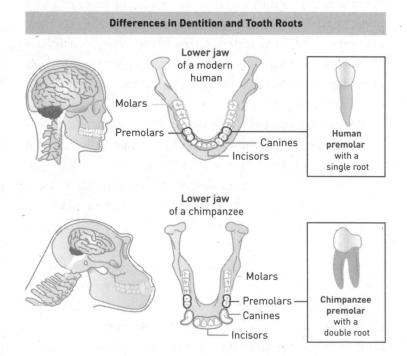

Differences in Dentition and Tooth Roots

Lower jaw
of a modern
human

Molars

Premolars

Canines

Incisors

Human
premolar
with a
single root

Lower jaw
of a chimpanzee

Molars

Premolars

Canines

Incisors

Chimpanzee
premolar
with a
double root

When von Freyberg and von Koenigswald examined the lower jawbone of *Graecopithecus*, they had to rely on what they could see. Back then, it was impossible to look at the inside of a fossil without destroying it. Today, using computed tomography (CT) scans that are accurate to the micrometer, you can X-ray a fossil and make the invisible visible. The process is the same as the one used by medical doctors, except in our case it is being used to examine the remains of dead life-forms up close. Micro CT scans make it possible to combine thousands of cross-sectional images to create three dimensional representations of hidden structures that are accurate to a few thousandths of a millimeter.

In the spring of 2015, my team and I were beyond excited when we X-rayed the lower jawbone of *Graecopithecus* using this high-resolution technique. The result was a complete surprise. We had expected to identify the typical features of a tooth belonging to a great ape. But what we saw was much more remarkable: the roots of the canines and molars of *Graecopithecus* were shortened. On the premolars, we even found that almost 50 percent of the roots had fused and the tips were converging, in other words, they were curving inward.

Graecopithecus also had fewer root canals compared with great apes living today and great ape fossils. The premolar from Bulgaria shared all these characteristics.

It was difficult to believe, but when you added all this together, the tooth from Bulgaria and the teeth from *Graecopithecus* did not look much like great ape teeth at all. They more closely resembled the teeth of a few early hominin species from the genera *Ardipithecus* and *Australopithecus* that had been found in Africa. The famous African finds were between 5.8 million and 2 million years old. But when exactly had *Graecopithecus* lived?

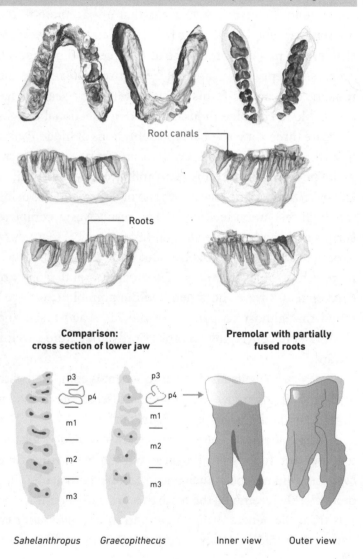

CT-Scan of the Lower Jawbone of *Graecopithecus freybergi*

Root canals

Roots

Comparison:
cross section of lower jaw

Premolar with partially
fused roots

p3
p4
m1
m2
m3

p3
p4
m1
m2
m3

Sahelanthropus   *Graecopithecus*   Inner view   Outer view

Dating Using Fossils to Tell Which Way Was Up

Geologists today use a variety of different physical measurements to date fossils. One of them, magnetostratigraphy, uses information from the Earth's magnetic field. Measuring the direction of magnetic particles in rocks has shown that the Earth's magnetic field reverses itself at irregular intervals. That means that although today the magnetic North Pole is close to the geographic North Pole, the magnetic North Pole can switch to the geographic South Pole. The last of these switches in polarity happened about 800,000 years ago. There were many more before then, and the exact date of each change in polarity is well known from information stored in sediments.

This kind of paleomagnetic information can be read from rocks using a magnetometer, an instrument that measures magnetization. It is more difficult to read this information from fossils, because whoever examines the fossils must know not only the exact geographic coordinates of the site where the fossils were found but also the fossils' orientation in the ground—that is to say, which side was up and which was down. A thorough documentation of how the samples were collected is essential. If you have that, then you have a kind of "fossil level"—a bit like the bubble level carpenters use to make sure their work is aligned correctly—that shows the direction of the gravitational pull when the fossils were deposited.

We were lucky, because once again, the other finds from Pyrgos proved to be invaluable. Among the more than thirty bones stored in Nuremberg, we came across two large, cracked metatarsal bones from the foot of a giraffe. These hollow bones were half-filled with sediment. Crystals of calcite had formed above the sediment, as they had done in the finds from Pikermi that the Bavarian soldier thought were filled with diamonds back in 1838.

The Earth's Magnetic Field and the Principles of Paleomagnetism

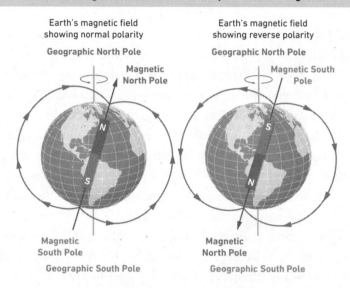

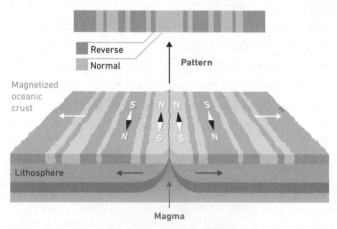

rises and cools. The prevailing direction of the Earth's magnetic field is encoded in rocks as they form.

The alignment of the sediment surface gave us the information we needed to reconstruct the exact orientation of the giraffe bones in the sediment. Using these fossil levels, we could deduce the Earth's magnetic field at the time the fossils in Pyrgos were deposited.

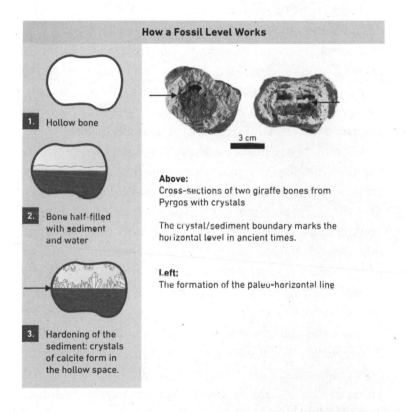

How a Fossil Level Works

1. Hollow bone

2. Bone half-filled with sediment and water

3. Hardening of the sediment: crystals of calcite form in the hollow space.

3 cm

Above:
Cross-sections of two giraffe bones from Pyrgos with crystals

The crystal/sediment boundary marks the horizontal level in ancient times.

Left:
The formation of the paleo-horizontal line

We combined the paleomagnetic data from Pyrgos, Pikermi, and Azmaka with data harvested from three other chronological techniques. We could then date the *Graecopithecus* fossils and the other Pyrgos finds fairly accurately: they were 7.175 million years old.[9] The premolar from Bulgaria

was about sixty thousand years older. That is around the time
when genetic data show that the human evolutionary line
had in all probability already split from chimpanzees.[10] We
calculated that the fossils from Pikermi were about 7.3 million years old, and we discovered that they came from eight
different layers of sediment that had been deposited over a
period of forty thousand years.

Dietrich and von Freyberg, therefore, had not been that
far off when they supposed that El Graeco belonged to the
same world as the fossils at Pikermi. We could also confirm
Gustav Heinrich Ralph von Koenigswald's stated opinion that
El Graeco was slightly younger than the Pikermi fossils.

What these scientists could not have imagined, however,
was the evolutionary significance of the find. The new analyses proved that *Graecopithecus freybergi* was significantly older
than the oldest potential early hominin from Africa. (In this
book, for simplicity's sake, we call all humanlike creatures
that evolved after the split from chimpanzees and before the
genus *Homo*—that is, that evolved before humans—"early
hominins," and we call the extinct members of the human
genus "early humans.")

The dates confirmed what the teeth had already told us:
Graecopithecus and the owner of the tooth found in Bulgaria
potentially stood right at the beginning of the human evolutionary line. Might El Graeco be the long-sought-after
original ancestor of humans? The fact that he came from
Europe added to the luster of his discovery and challenged
traditional assumptions about human evolution—first and
foremost the assumption that the early evolution of humans
had taken place exclusively in Africa. And this was not the

only find that suggested some rethinking might be required. In order to make an informed judgment here, let us take a look at how the search for the origins of humankind began and how it has progressed to today.

Part 2

THE REAL
PLANET
OF THE APES

- 6 -

DISASTERS AND
SUCCESSES

A Short History of the Search for Our Origins

Q UESTIONS ABOUT OUR origins are probably as old
as humanity itself. Where do we come from? Who
were our original ancestors? Why did we evolve in
the first place? And what made and makes us
the kind of creature we are?

People have sought answers to these questions for a long
time, mostly through religion and philosophy. It was only
with the rise of the natural sciences that scientific answers
to these basic questions gradually came to the fore, answers
based on careful observations, detailed measurements, and
increasingly refined analytical methods.

However, paleoanthropology, the relatively young science
of prehistoric humans and their ancestors, has not always
been characterized by goal-oriented, self-critical research.
It has also had its share of accidental discoveries, vanity-
driven agendas, dazzling personalities, and unscrupulous

frauds. Paleoanthropologists, like paleontologists and archeologists, are scientists armed with shovels. Their reputation as treasure seekers or bounty hunters precedes them, and the circumstances under which some historically important finds came to light seem faintly comical even today. Other episodes are so ludicrous that when we look back on them knowing what we know today, it's hard to believe they ever happened. And yet, over time an increasingly detailed picture of our complex evolutionary history has come into focus.

The journey to the beginnings of this fascinating evolutionary story of humans and great apes, the highlights of which we will touch on here, takes us back to nineteenth-century France. In 1856, the French paleontologist Édouard Lartet[1] received a parcel containing perplexing contents: the fragments of the fossil of a lower jawbone and a humerus bone from a clay pit near Saint-Gaudens in southern France. The person who had sent the package was a Mr. Fontan. Workers had given him the bones, and now he wanted an expert opinion about what they were.

Lartet, one of the foremost fossil specialists of his time, quickly identified the bones as being the remains of an extinct great ape. He described the find in a scientific article and gave it the name *Dryopithecus fontani*. *Drys* in ancient Greek means "oak" and *pithecus* means "ape." Because fossil imprints of oak leaves had also been found at this site, Lartet concluded that this ape must once have lived in an oak forest. He then went a step further. He compared the *Dryopithecus* remains with chimpanzee bones and concluded that the fossil was a link between great apes and humans. And with that, the "ape from the oak forest" became the first piece of the puzzle in a scientific conversation about our roots that continues to this day. It was the first of a series of discoveries that eventually spread way beyond Eurasia.

Early Human or Misshapen Cossack?

At first, Europe remained the focal point of the research, because, in a happy coincidence, the same year *Dryopithecus* was first described another fascinating find was unearthed. And once again, it was found in a quarry. Italian workers were preparing a cave in the Düssel valley in Germany for limestone mining when, in the hardened clay layer on the cave floor, they came across bones. At first, the men thought nothing of their find. They were always coming across the remains of animals that had sought shelter in caves, and so they threw the fossils on a spoil pile. That was where Wilhelm Beckershoff, the owner of the quarry, found them. Because he thought they were the remains of an extinct cave bear,[2] he passed them on to a naturalist named Johann Carl Fuhlrott—and inadvertently drafted a chapter in scientific history. Fuhlrott quickly realized that they were human bones, but when he examined them, he noticed some differences between them and the bones of modern humans. One fossil in particular drew his attention: a relatively flat skull cap with a massive, primitive-looking ridge of bone that bulged out over the eye sockets.

In the end, he concluded that the finds must belong to a human "from prehistoric times." He and a professor of anatomy named Hermann Schaaffhausen presented the find to the scientific community in 1857, but no one took them seriously. For example, some people suggested they had found the remains of a Russian Cossack who had sought refuge in the cave during the Napoleonic Wars and his bones had been deformed as a result of illness. A broken arm that had healed badly must have caused the poor man so much pain that he furrowed his brow constantly, which made the bones above his eyes bulge.

It took many years, during which time further comparable finds were made and British paleontologists interceded, before it became widely accepted that these fossils really did belong to an early form of human. Because the section of the valley where the Italian stone workers had discovered the fossil was called the Neander valley (*Neandertal* in German), the fossil was retroactively called *Homo neanderthalensis*.[3]

Lemuria, Navel of the World

In this early phase of paleontology, therefore, much of the evidence suggested that the cradle of humanity lay in Europe. But although the number of finds was still very small, some experts were already developing new ideas. As early as 1863, the British biologist Thomas Henry Huxley wrote in his book *Evidence as to Man's Place in Nature* that the true source of humanity could lie in Africa, because he recognized that the great apes living there were our closest living relatives.

Huxley was a contemporary of Charles Darwin. The two men knew each other, but Huxley, like many scientists of his day, stood in Darwin's shadow after Darwin published his seminal work, *On the Origin of Species*, in 1859. The first time Darwin offered an opinion on the ancestry of humans was in 1871 in his book *The Descent of Man, and Selection in Relation to Sex*. In it, he agreed in principle with Huxley's idea that

Africa was the cradle of humanity, but he added, in relation to Lartet's *Dryopithecus*, "But it is useless to speculate on this subject; for two or three anthropomorphous apes, one the Dryopithecus of Lartet... existed in Europe during the Miocene age;[4] and since so remote a period the earth has certainly undergone many great revolutions, and there has been ample time for migration on the largest scale."[5] So even Darwin recognized that migrations and complex geography could also have left their mark on human evolution, something that is often overlooked when people read his writings today.

Europe and Africa were not the only places being considered in those years as possible cradles of humanity, though. The German evolutionary biologist Ernst Haeckel, an ardent proponent of Darwinism, compared the anatomy of orang-utans, gibbons, and humans and concluded in his 1868 book *Natürliche Schöpfungsgeschichte* (published in English in 1884 as *The History of Creation*) that humans most closely resembled these great apes and therefore probably arose in southern Asia. At the time, this was a revolutionary idea. He even went so far as to give his alleged Asiatic early human a name, *Pithecanthropus primigenius*, which means "original ape man."

Haeckel was also a proponent of the idea that a sunken continent in the Indian Ocean called Lemuria had once formed a land bridge between East Africa and Southeast Asia. On this continent, he believed, early great apes had evolved into the ancestors of humans and today's great apes. Working from mainstream geological thinking at the time, he thought the Indonesian islands of Java, Sumatra, and Borneo were what remained of this continent. Today this idea is considered absurd, but Haeckel was one of the most influential natural scientists of his day, and his books were bestsellers.

The Discovery of Java Man

One of Haeckel's followers was a Dutch doctor and anthropologist named Eugène Dubois. Unlike Haeckel, Dubois was convinced that in addition to anatomical studies of living animals and humans, fossils were needed to reconstruct the evolutionary history of humankind. And so, in 1886, he resigned from his position as a professor of anatomy at the University of Amsterdam and traveled to Sumatra as an army doctor so he could search for the fossils of early humans. What happened next is an amazing story of perseverance and luck. In his first years there, he did indeed discover numerous fossils—but not a single trace of the link he was looking for between great apes and humans.

Dubois did not give up, however, and in 1890, he moved to Java and settled there permanently. Heat, malaria, catastrophically awful sanitation—none of that could deter him from pursuing his goal. Finally, in 1891, his assistants came across a molar on the banks of the Solo River. In the sediment only a few feet away was a primitive-looking flat skull cap that, like the Neanderthal skull cap, had pronounced bony ridges above the eyes. Electrified by his success, Dubois continued to search, and in 1892 he found a human-like thighbone at the same site. He was soon convinced that he had finally found the "missing link."

In 1894, he published his find with the name *Pithecanthropus erectus* ("upright-walking ape man"). But Dubois, too, was initially hugely disappointed. The academic world treated him the same way it had treated Johann Carl Fuhlrott, and his colleagues declined to give him the recognition he was hoping for. And, once again, the arguments used to discredit the find bordered on the absurd. One, for instance, suggested that the bones were nothing more than the remains of an extinct giant gibbon.

It was not until the 1920s and 1930s, when further comparable fossils were found first in China and then in Indonesia, that the scientific community admitted that Dubois had been right. In the 1950s, *Pithecanthropus erectus* was finally renamed *Homo erectus* (which translates as "upright man"). Since then, it has been confirmed that Dubois's finds were about 1.5 million years old. His find, commonly referred to as Java Man, is now the type specimen—the example that defines the species—of *Homo erectus*, and Dubois was posthumously considered to be one of the most important early paleoanthropologists.

Many early scientists, with their Western points of view, were not comfortable with the thought that humans originated in Asia, and so they were relieved when, at the beginning of the twentieth century, new finds refocused attention on Europe. In 1907, Daniel Hartmann, a worker at a sandpit near Heidelberg, Germany, stumbled across a well-preserved lower jawbone complete with teeth that looked as though it might have come from a primitive human. By then, the sandpit had been closely monitored for going on twenty years to check for fossils that had been unearthed as the sand was dug. The person responsible for overseeing fossil excavations at the sandpit was the paleontologist Otto Schoetensack, and the first description of this new species of human was the crowning achievement in his scientific career. In honor of his chosen home, Heidelberg, Schoetensack called the find *Homo heidelbergensis*.[6] Today, most experts classify *Homo heidelbergensis* as an evolutionary precursor to Neanderthals and modern humans.[7]

Skilled Craftsmanship and the Drive to Deceive

In the years that followed, however, a find from Great Britain attracted much more attention than the discovery of *Homo heidelbergensis*. In 1912, Charles Dawson, an amateur geologist, and Arthur Smith Woodward, the curator of the geology department at the prestigious British Museum of Natural History, publicly presented a skull with a lower jawbone that showed a combination of human and ape-like features that had never been seen before. Dawson and Smith Woodward indicated the find came from a gravel pit near Piltdown in southeastern England. They said they had filled in the missing pieces to give a better idea of what this unknown human ancestor would have looked like. The find immediately attracted a great deal of attention, partly because Dawson and Smith Woodward had chosen the name *Eoanthropus dawsoni* ("Dawson's dawn man") and with that, they intentionally placed the fossil at the beginning of the human evolutionary story.

British experts in particular were impressed, because Piltdown Man, as he was also called, did indeed seem to be older than the Neanderthals and *Homo erectus*. Moreover, the skull with its unique combination of a large brain with an apelike lower jaw seemed to confirm a widely held opinion of the times that the evolution of a large brain was a necessary precondition for the development of an upright gait, the ability to make tools, and human speech. Critical voices that pointed to the fact that the find with its remarkable combination of primitive and modern features did not fit with other fossils were ignored. It was not until 1953 that Piltdown Man was revealed to be one of the greatest hoaxes ever carried out in the history of science.

With a definite intent to deceive and a high level of skill, the lower jaw of an orangutan had been attached to a human skull from the Middle Ages. The originator of the hoax had colored the bones to make them look older. To further conceal the deception, they had also filed down the teeth and broken off bits of bone where the jaw attached to the skull, which might otherwise have given the game away. The counterfeiter successfully led even experienced experts to believe that it was an authentic fossil.

It was never clearly proven who had made Piltdown Man. Both "discoverers" died before the hoax was revealed, Dawson in 1916 and Smith Woodward in 1944. Today Dawson is the prime suspect. Genetic analyses carried out in 2016 provided the final proof that Piltdown Man was a chimera. Today, people suspect the deception went undetected for so long because patriotic fervor clouded the judgment of many experts. After the Neanderthal finds in Germany, Belgium, and France, British scientists were hoping for their very own "First Briton," which was probably why they did not scrutinize the find more closely.

In hindsight, an alleged discovery by the American paleontologist Henry Fairfield Osborn was similarly ludicrous. In 1922, he classified a tooth found years before by a farmer in Nebraska as a fossil from a human ancestor. With nothing more than this tooth to go on, he brought North America to the table as the cradle of humanity. From the beginning, most other experts rejected Osborn's *Hesperopithecus haroldcookii* ("ape of the Western world discovered by Harold Cook"). In 1927, he admitted that he had been mistaken. Further investigation had shown the fossil to be the severely weathered tooth of an extinct species of pig, further remains of which had been found at the same site. And so Nebraska Man faded from the spotlight after just a few years.

AT THE BEGINNING of the twentieth century, therefore, the situation in paleoanthropology was already somewhat confusing. Numerous researchers had collected fossils from an enormous region—extending from Western Europe to Indonesia—and each fossil was allegedly proof that humans had arisen where it had been found. The only place where no finds had been made by that time was Africa.

An Awakening in Africa and a Breakthrough in Asia

The African fossil drought ended abruptly in 1924, when a worker in a limestone quarry near the South African village of Taung came across a fossil skull. The find was handed over to the anatomist Raymond Dart at the University of the Witwatersrand in Johannesburg, who quickly realized its importance and, in 1925, described it as the fossil of an early hominin child. Dart called it *Australopithecus africanus* ("southern ape from Africa"). But, like many other pioneers in paleontology who had described new finds, he initially encountered massive resistance to the name. People accused him of not noticing that the skull bore a strong resemblance to the skull of a young gorilla or chimpanzee and that *Australopithecus* was likely more closely related to them than to humans.[8] One further feature of the fossil raised the ire of critics. A natural stone cast had formed inside the skull, a kind of fossilized impression of the brain. This showed that a mature *Australopithecus africanus* would have had a brain volume of about 440 cubic centimeters.[9] That is slightly larger than the brain of a modern chimpanzee. Dart's critics completely dismissed the possibility that a human ancestor could have had such a small brain. And so it was not until 1947 that his classification of *Australopithecus africanus* as an early hominin was finally recognized.[10] Today the fossil is famous around the world as the Taung Child. Dating has shown the skull to be between 3 million and 2.5 million years old.

Between the two world wars, scientists once again found fascinating new *Homo erectus* fossils in Asia, including the famous Peking Man, the name given to a large number of *Homo erectus* fossils from limestone caves near Peking, including 150 teeth and fragments from 14 skulls. These fossils were unearthed during numerous digs in the 1920s and 1930s. Endangered by the turmoil caused by World War II, the finds were supposed to be shipped from China to the United States in 1941, but they disappeared when Japanese troops seized the train that was transporting them to the harbor. Except for a few teeth and individual bones that were discovered in digs carried out later, nothing remains of this rich collection but the drawings and casts made of the original fossils. Modern analysis methods have shown that the finds are about 780,000 years old.

After the discovery of this second site in Asia, that continent was once again considered a potential hot spot for human evolution, especially as *Homo erectus*, like the Neanderthals, looked much more like a human than *Australopithecus africanus* did. That changed in the late 1950s when, for the first time, sensational finds came to light in East Africa. These finds continue to shape our ideas about human evolution to this day, and they are inextricably entwined with a British family from Kenya that has produced three generations of famous paleoanthropologists: the Leakeys.

The Leakey Dynasty

The British anthropologist Louis Leakey and his wife, Mary, had been looking for fossils and artifacts in the Olduvai Gorge, a barren, remote canyon on the edge of the Serengeti in the north of what is now Tanzania, since the 1930s. The canyon, which was carved out millions of years ago by runoff from torrential rains, was part of the Great Rift Valley. In July 1959, Mary Leakey found an almost completely intact skull there that she identified as an early hominin. A month later, the Leakeys published the find under the name *Zinjanthropus boisei*, soon referred to by the Leakeys as Zinj. *Zinj* is a local term for the area where the find was made and *boisei* is a reference to Charles Boise, who was funding the Leakeys' research.

Right away you could see that this skull was completely different from that of the Taung Child. A massive bony crest on the top of the skull was particularly striking. Powerful muscles would have been attached there, which would have allowed Zinj to chew food that was particularly hard and tough. The crest earned him the popular nickname Nutcracker Man. Later the find was discovered to be "only" a parallel branch to the human evolutionary line and it was therefore renamed *Paranthropus boisei*, "Boisean human neighbor." For the Leakeys, however, the Zinj find was a breakthrough.

A few months after this discovery, the Leakeys' digging team made another find in the Olduvai Gorge. It turned out to be one of the most important finds ever made in Africa: *Homo habilis*. All they found at first was a jaw fragment with a wisdom tooth still in it. Then, close by, they found an almost intact lower jawbone. The Leakeys' most recent find attracted attention from around the world. The name they chose, *Homo habilis*, meant "gifted" or "skilled" human, often translated as Handy Man.[11]

For the first time, paleoanthropologists had an African fossil classified as belonging to the genus *Homo* rather than the early hominin species *Australopithecus* and *Paranthropus*. The basis for this classification was the very simple stone tools the Leakeys found at the site. They were sure that only *Homo habilis* could have fashioned these rocks into hammering and cutting tools.[12] The researchers guessed that these early humans used the tools to smash animal bones, which contain nutritious marrow, or to crack open particularly hard fruits. Assuming that the capability to make and use tools marked the transition from early hominin to the human genus, *Homo habilis* was declared to be the oldest "true" human.[13]

In 1978, Mary Leakey arrived at another milestone. About 30 miles (50 kilometers) south of the Olduvai Gorge, in a place called Laetoli, her team came across fossilized footprints of early hominins who walked upright. The footprints were at least 3.6 million years old. At the time, the footprints were the oldest proof of upright gait in the world. But what kind of creature had walked across rain-soaked volcanic ash all those years ago?

One possible answer came with the publication in that same year of a find that is considered to this day to be the most important discovery ever made in paleoanthropology: Lucy, a 3.2-million-year-old partial skeleton of an early hominin. An international research team under the leadership of the famous paleoanthropologist Donald Johanson had found her bones in the Afar Region of Ethiopia. Lucy became known as *Australopithecus afarensis* ("southern ape from Afar").

Lucy's skeleton was 40 percent complete, a sensation in a research discipline that previously had had to rely on individual bones or teeth to reconstruct a story of evolution that was millions of years old. This much was clear: Lucy was only 3 feet (1 meter) tall, weighed less than 66 pounds (30 kilograms), and could walk upright.[14] She got her name from the Beatles' song "Lucy in the Sky With Diamonds," which was playing in the research camp when she was discovered.

With every new find, the idea that humans had evolved only in Africa gained currency. Another groundbreaking find came to light in 1984. That was when one of the people working with Richard Leakey, one of Louis and Mary Leakey's sons, found fascinating new fossils of *Homo erectus*. *Homo erectus* had already been discovered in Africa, but this time the find proved to be outstanding.[15] Researchers had worked for many years in the area around Lake Turkana, sifting through 1,500 tons of sediment and eventually exposing about 90 percent of a human skeleton that was about 1.5 million years old—one of the most complete paleoanthropological finds of all time. From the bones, scientists could prove that these were the remains of a boy about nine years old who would likely have been close to 6 feet (1.80 meters) tall had he lived to be an adult. They called him Turkana Boy.

Taung in South Africa, the Olduvai Gorge in Tanzania, the Afar Region in Ethiopia, and Lake Turkana in Kenya are to this day the regions that have contributed the best-known chapters in the story of human evolution. These discoveries convinced most scientists that humans and their ancestors evolved in Africa and that the first *Homo erectus* left this continent and advanced as far as Asia. The fact that the oldest fossils of modern humans, *Homo sapiens*, were also found in Africa confirmed this narrative in the eyes of almost all paleontologists.

Our Oldest Original Ancestors

But it's worth taking a closer look. Recent studies based on genetics and molecular biology have narrowed the time when humans and chimpanzees split to between 13 million and 7 million years ago. It follows that the fossils of our oldest ancestors should also date from this time. But despite their tremendous importance, all the African finds are a few million years younger than this.

Moreover, fossils from the last common ancestors of humans and great apes living today have never been found in Africa and yet, given the current thinking, this is exactly where you would expect to find them. For exactly the time period I have just mentioned, there is clearly a gaping hole in the African fossil record when it comes to the ancestors of great apes.

In Eurasia, however, researchers have found a great number of such fossils from this phase of evolution. Shouldn't the cradle of humanity lie where the oldest bones have been discovered? Did the direct ancestors of chimpanzees and gorillas perhaps not come out of Africa after all? Could it be that they migrated there later? *Graecopithecus* points to this way of thinking. Those who support the current school of thought have so far categorically rejected such suggestions. Discoveries from Eurasia are declared to be side branches of evolution at best, evolutionary dead ends that made no contribution to the evolution of present-day great apes and certainly not to humans—either that or nothing is said about them at all.

Back in 1992, my Canadian colleague David Begun published a comprehensive analysis of all the great apes known at that time and came to the remarkable conclusion that the fossil great apes of Europe were indeed the starting point for all the African great apes and for humans. To this day, his conclusion is considered to be an isolated opinion, even though many new finds have been added since then, as we will see later in this book.

Usually the true significance of a find becomes clear only years later with the benefit of hindsight. Most importantly, later fossil digs can shed new light on earlier discoveries. Knowing where to look in order to make such discoveries, in which regions of the world—in which rock layers of which geological age—is just as important as patience and persistence. And when you also get the tiniest bit of luck when you dig in exactly the right spot, you can make the most amazing finds.

But before we get to these recent discoveries, let's turn our attention to the early phases of great ape evolution.

- 7 -

AFRICAN
BEGINNINGS

The Golden Age of Ape Evolution

W HETHER WE LIKE it or not, from a biological point
of view we are without a doubt apes, two-legged
apes with relatively large brains. Apes and monkeys,
known in scientific terms as primates, have developed a great
diversity in appearance, size, and features over the course of
their evolution. Their ancestors lived over 60 million years
ago. To understand their evolutionary history, you need to
know some basic terminology and also how the primate line
is subdivided.

Biologists make a clear distinction between Old World
anthropoids and New World monkeys. The latter played no
role in human evolution. They lived, and still live, in Central and South America. The Old World anthropoids, on the
other hand, are critical to our evolutionary story. They are
divided into Old World monkeys, which have tails, and apes,
which do not. Apes (Hominoidea) are further subdivided

into lesser apes—gibbons and siamangs (Hylobatidae)—and great apes (Hominidae). Great apes include all great apes living today and humans and their extinct ancestors. First, the orangutans (Ponginae) split from the lineage leading to gorillas, chimpanzees, and humans (Homininae). Then the gorillas (Gorillini) split off, and finally, the chimpanzees and bonobos (Panini), leaving the human line (Hominini). Over millions of years, early hominins (prehumans) evolved into species of early *Homo* (early humans), and ultimately into us: *Homo sapiens.*

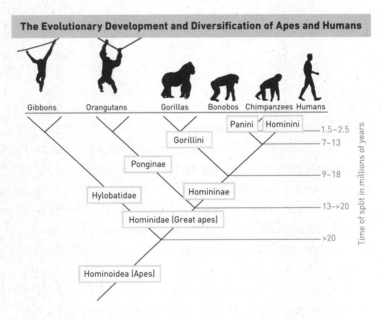

The search for our roots, therefore, reaches way back into the evolutionary history of primates. The changes in body shape of the different species, which evolved as they settled in new habitats and adapted to changes in climate, are the cornerstones of human evolutionary history.

It is astounding how many genes we share with our closest living relatives. The genetic material of humans and chimpanzees is 98.7 percent identical. We share 98.3 percent of the DNA sequences in the nuclei of our cells with gorillas and 96.6 percent with orangutans. Chimpanzees and bonobos are more closely related to humans than they are to gorillas.

As mentioned earlier, genetic analyses[16] suggest that our oldest ancestors split from chimpanzees as far back as 13 million years ago, and each lineage has been following its own evolutionary path for the millions of years that have passed since then. This means that the human lineage did not evolve from chimpanzees, but that humans and chimpanzees are, to put it in simple terms, sister species. That in turn means that both lineages go back to a single as-yet-unknown species that in all probability did not bear much resemblance to modern chimpanzees let alone to modern humans.

Today we know of about one hundred species of fossil and living great apes. We can divide their history into three phases, each lasting for about 7 million years. The oldest ape or nearly ape species were found in Africa and lived about 21 million to 14 million years ago. The best glimpse we have into the diversity and habitat of these ancestral apes is provided by 18-million-year-old fossils from Rusinga Island in Lake Victoria in Kenya. Paleontologists have been digging there for the past one hundred years, and their finds have created a detailed picture of a lost world.

Apes Among Strange Creatures

Eighteen million to 14 million years ago, Rusinga Island was located about 340 miles (550 kilometers) south of the equator. The climate was tropical. Dry periods were inter-spersed with monsoon rains, and thick forests covered the land. The first primitive apes shared the island with animals we still have today and with exotic creatures that died out long ago. Monitor lizards grazed their way through the under-growth, while chameleons hunted insects in the branches overhead. Bizarre-looking animals such as giant elephant shrews—relatives of aardvarks and golden moles with long legs and an elongated nose that looked like a trunk—scoured the undergrowth for beetles, ants, and millipedes.

The strangest creature in these forests, however, was an 8-foot-tall (2.5-meter) "hooked beast" or chalicothere. At first glance, chalicotheres look like the result of a joke played by nature. Related to tapirs and rhinoceroses, they had extremely long front legs and short, stubby hind legs. Thanks to their odd body shape, chalicotheres had no diffi-culty rearing up on their hind legs to reach fresh green growth on branches and twigs high above their heads. They grabbed hold of branches with their long, inward-curving claws and then used them like rakes to efficiently strip off the leaves.

At that time, the average temperature of the Earth was about 14.4 degrees Fahrenheit (8 degrees Centigrade) higher than it is today. Huge clouds of steam hung in the air, trapping heat like a greenhouse gas, and the level of carbon dioxide in the atmosphere was 50 percent higher than it is today.[17] This phase, the warmest phase in the last 30 million years,[18] is called the Miocene Climatic Optimum.

Almost one-third of the apes we know of—about thirty species—come from this early phase of great ape evolution. Together they tell the story of the first Golden Age of the apes, as David Begun has called it.[19] Not all finds of these archaic species have been verified. For some, only a few teeth or a jawbone exist. For others, however, parts of the skeleton have also been unearthed.

The most important genus is *Ekembo*, two species of which were discovered on Rusinga. Members of this genus were leaf eaters and weighed from 22 pounds to 110 pounds (10 kilograms to 50 kilograms). At first glance, these archaic apes looked like Old World monkeys, of which there are about 138 species today. The best known are baboons and macaques. *Ekembo*, however, lacked a tail. In this genus, just a few bones, curved forward and grown together to form a single tailbone, represent the vestiges of the tail. The lack of a tail is a distinguishing feature of all apes, and it meant *Ekembo* was missing a crucial aid when balancing on branches or leaping through the trees. That in turn meant that *Ekembo*'s hands and feet were more important for steering and holding on, as the bones of its fingers and toes clearly show. *Ekembo* certainly had an extremely firm grip.

Apart from these few anatomical changes marking the transition from monkey to ape, *Ekembo*'s body shape was still quite primitive in comparison to today's great apes. Its arms and legs were of equal length, it could not completely straighten its elbows, and its fingers and toes were straight, which meant that when it walked on all fours, it placed both its hands and its feet flat on branches or on the ground. This is not how apes walk, but it is still typical of the way many monkeys walk today. You can clearly see this, for example, if you watch baboons moving around in zoos.

About 1 million years after the first *Ekembo* appeared, another primitive ape came on the scene, and many species of this ape were also discovered in Kenya. Its name was

Afropithecus, and its most striking feature was powerful jaws for chewing. From the skull fragments that have been found, it is clear that its chewing muscles were much stronger than those of *Ekembo*. In addition, *Afropithecus* had an extra thick layer of tooth enamel, a first in the evolution of apes. Usually forest-dwelling monkeys and apes that eat fruit and leaves have tooth enamel that is only about 0.02 inches (0.5 millimeters) thick, because their food is soft and causes very little wear on the teeth. Building tooth enamel is an energy-intensive process, so no mammal has more than it needs. *Afropithecus*'s thick enamel, therefore, points to a change in diet. It was probably eating food that was harder and tougher.

Finds very similar to *Afropithecus* have also been discovered in the desert in Saudi Arabia. They prove that even in those days, apes inhabited a wide area. Saudi Arabia back then lay along a long, narrow stretch of sea that connected what was later to be the Mediterranean with the Indian Ocean. During this time, 17 million to 16 million years ago, there were extensive forests there as well.

Clearly the narrow stretch of ocean was not an insurmountable barrier in those days, because that was when primitive apes made it to the European landmass for the very first time, as proven by a single molar with thick enamel found in 1973 in Engelsweis near Signaringen in southern Germany. In 2011, my team and I calculated that the layers of limestone where this tooth was found were 15.9 million years old. Although the tooth could not be classified definitively as belonging to *Afropithecus*, it did prove that even at that time apes—even if it might have been only a few individuals and even if they apparently did not stay long—were extending their range far to the north. This hypothesis is supported by fossils from other animals that we know migrated.

The Advance of Africa

For the past 100 million years, the African continental plate has been getting closer and closer to Europe and Asia. Fourteen million years ago, the sea that separated the Arabian Peninsula from the Eurasian landmass was shallow, and minor fluctuations in its level were enough to reveal small island chains and create land bridges.

In the tropical temperatures of the Miocene Climatic Optimum, forests of palm, ebony, and mahogany spread over vast parts of Europe.[20] Coasts were edged with tangled bands of mangroves. Three different species of crocodiles swam in the rivers.[21] The largest of the three, *Gavialosuchus*, grew up to 23 feet (7 meters) long. The air was so moist that snakehead fish traveled overland from one lake to the next.[22] We do not know why the ape found at Engelsweis did not settle in Europe back then despite the favorable climate. It could well be that we have simply not yet found its descendants.

Many fossil finds prove that the ecological histories of Africa and Europe are inextricably intertwined, and together they document the story of a slow convergence. At a speed of 0.04 inches (1 millimeter) a year, the African continental plate continues to move north even today. When its northward drift began, both continents were still separated by the 2,480-mile-wide (4,000-kilometer) ancient Tethys Ocean. Since then, Africa has drifted 620 miles (1,000 kilometers) north, and the southernmost coast of Europe has extended in the direction of Africa. A 1,250-mile (2,000-kilometer) stretch of the southern edge of Europe rose as the ocean floor of Tethys slid beneath it, throwing up the European mountain ranges of the Pyrenees, the Apennines, the Alps, the Dinarides, and the Balkans. The Old World, as seafarers called Afro-Eurasia at the beginning of the modern era, was coming together. And in a future only a few million years away from our present, the Mediterranean will eventually disappear completely, buried under new mountain ranges, and a new supercontinent will be formed.

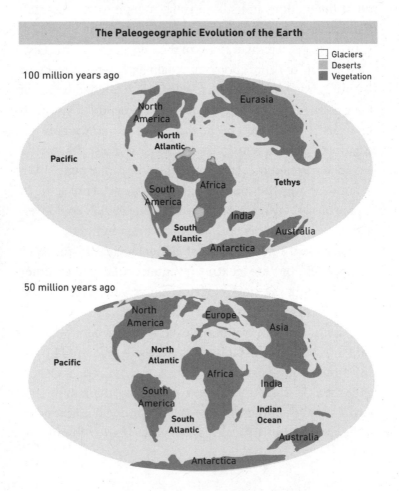

The Paleogeographic Evolution of the Earth

14 million years ago

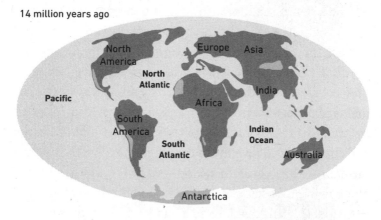

18,000 years ago
(the last glacial period)

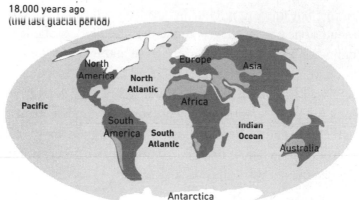

50 million years
into the future

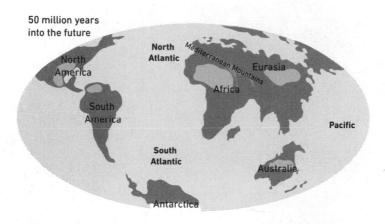

The story of the diminishing gap between the two continents is also the story of the reciprocal migration of fauna.
Even before any land bridges were formed, many animals
used the islands in the Tethys Ocean as stepping stones so
they could expand their ranges. First, primitive hoofed animals and primates[23] traveled from Eurasia to Africa, followed
by primitive ruminants, pigs, rhinoceroses, monitor lizards,
and chameleons. Traveling in the other direction, from Africa
on their way to settle in Europe, were the first true frogs,
along with proboscideans, monkeys, and primeval animals
with double horns that looked a bit like rhinoceroses.[24]

The first temporary land bridge between the three continents appeared in what is today the Middle East about 17
million years ago. But by that time the mammals of the Old
World were already relatively uniformly distributed, and they
were well mixed when, 13.8 million years ago, the Tethys
Ocean began a complete retreat from the Arabian region,
marking the end of the first phase of ape evolution. From that
time on, a new chapter in their evolutionary story began, and
its main theater was Eurasia.

PROGRESS IN EUROPE

Great Apes in Oak Forests

A S HAPPENS SO often in the story of the evolution of life, the second phase of great ape evolution, between 14 million and 7 million years ago, was ushered in by climate change. The Miocene Climatic Optimum ended about 14 million years ago, and eastern Antarctica froze over completely. One factor in this was that a stable current that had developed around the South Pole was bringing cold water from far down in the depths up to the surface.[25] There was so much cold water that it absorbed and stored large quantities of carbon dioxide from the air. Another factor was that eroded rock was absorbing enormous amounts of greenhouse gases. The weathering of large quantities of rock reduces the levels of greenhouse gases in the atmosphere, because during this long, slow, natural process, minerals from the eroded rock bind with carbon dioxide to form other chemical compounds.[26]

The upshot was that the amount of carbon dioxide in the atmosphere was drastically reduced, causing long-term average temperatures to drop. There was about 20 percent more water frozen in glaciers than there is today, which had far-reaching consequences for the global climate and for the Earth's ecosystems. There was an average cooling of 9 degrees Fahrenheit (5 degrees Centigrade) and sea levels dropped by 165 feet (50 meters). Large areas of what had been shallow seas on the coasts of continents dried up. Africa and the Arabian Peninsula were now firmly joined to Eurasia via a land bridge.

Many mammals, including some apes, were increasingly taking advantage of the opportunity to expand their range north, and this time, the apes succeeded in settling down there for the long term. They were so successful that they were soon distributed over a large part of Eurasia—from the Iberian Peninsula to China. In many Eurasian forests, they quickly evolved to become the rulers of the canopy. It is not fantastic to talk of Earth at this time as being a Planet of the Apes. When apes conquered the northern latitudes, it was a milestone in the story of how we became human. If the apes had not reached Eurasia, humans would likely never have evolved, because the adaptations the apes made to survive in the different environment of northerly latitudes would never have been needed—and it was those adaptations that paved the way for humans to arise.

Édouard Lartet's *Dryopithecus fontani* ("the ape from the oak forest") that had been found in southern France belonged to one of the first genera of great apes that settled in Europe and evolved further while it was there. Lartet's find was just the first of many great ape fossils discovered on European soil. It likely weighed between 45 pounds and 90 pounds (20 kilograms and 40 kilograms), similar to a chimpanzee. The bones and teeth that have been found suggest that great apes in this genus lived in the tree canopy most of the time, where they fed on soft plant material that was easy to chew. They probably did not spend much time on the ground. Their faces were very similar to those of gorillas but much smaller.[27]

Since *Dryopithecus*, scientists have discovered eleven more genera of great apes from Europe and nine from Asia. They were found in Pakistan, India, Nepal, China, Myanmar, Thailand, France, Italy, Spain, Greece, Bulgaria, Turkey, Austria, Germany, Slovakia, and Hungary. Many of these finds already possessed very advanced features. For example, out of the numerous fossils that were found in an iron ore mine in Rudabánya in Hungary in the 1960s, researchers could build a picture of a great ape that had a brain that was comparable in size to that of a modern-day chimpanzee, although the ape was considerably slighter in build and weighed less. It was named *Rudapithecus hungaricus*. As the relationship between brain volume and body weight is a rough measure of the intelligence of an animal, *Rudapithecus* might even have been cleverer than the great apes alive today—with the exception of humans, of course. And intelligence, in turn, is proof of highly developed communication and social skills.

Between 2004 and 2009, a small region in Catalonia in northeastern Spain, Vallès Penèdes, made one international headline after another. A particularly large number of finds came to light near a village called Can Mata, where the nearby metropolis of Barcelona, with its population of millions, dumped its garbage. Many millions of cubic yards of soil were moved, leading to the discovery of thousands of fossils,[28] including three genera of great apes (*Dryopithecus, Anoiapithecus,* and *Pierolapithecus*) from the middle phase of great ape evolution.[29]

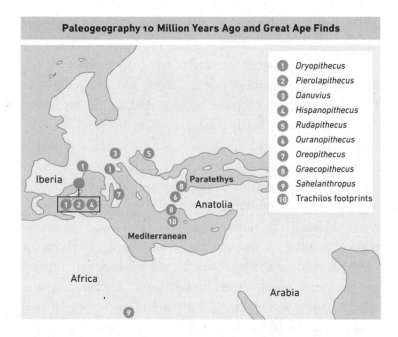

Fat Reserves for Lean Times

During the middle phase of great ape evolution, deciduous trees such as oaks and beeches were crowding out their tropical counterparts in many parts of Europe and Asia. At the same time, animals that flourished in the heat, such as crocodiles, snakehead fish, and chameleons, were also disappearing. It took time for the apes from Africa to adjust to the new vegetation of Eurasia, and they found themselves facing considerable challenges. For example, unlike in Africa, fruit here was not available year-round. Moreover, in the apes' new, more northerly home, there was far less sunshine in winter, and the days in that season were significantly shorter,[30] a circumstance that directly affected the supply of food.

In higher latitudes, day length rather than temperature dictates the formation and availability of buds and leaves. The plants sprout when the days get longer in the spring and drop their leaves when the days get shorter in the fall. This means that even back then, when winter days were warm but dark, most trees dropped their old leaves, did not grow any new ones until spring rolled around, and had no fruit on their branches in winter. Seasonal changes challenged the survival skills of tree-dwelling, leaf- and fruit-eating animals such as great apes.

There was another problem. Unlike trees in tropical forests, which grow three canopies, each 33 feet to 165 feet (10 meters to 50 meters) above the ground, the trees here formed a single low canopy. One reason for this was that most of the time there was less humidity in the air, which made the environment in the higher latitudes significantly drier than near the equator. The lack of moisture restricted the trees' ability to grow multiple canopies in their quest for light. The trees also did not grow so tall or so close together, and overall there were fewer treed areas. How was it, then, that the Eurasian great apes adapted so well to this difficult environment that they not only survived but thrived?

Molecular biology suggests a possible and highly plausible explanation. About 15 million years ago, Eurasian great apes experienced a genetic mutation that changed how they metabolized protein.[31] As a result of this mutation, their bodies could no longer produce the enzyme uricase. In most mammals, uricase is responsible for breaking down uric acid so it can be excreted in the urine. Without uricase, uric acid is more difficult to dispose of and so it builds up in the blood. That has serious consequences for both functional capacity and health. When there are elevated levels of uric acid in the blood, the body converts fructose, a sugar contained in fruit, into fat more quickly and stores it in the form of fat cells in the liver and in tissue. We humans have inherited this mutation from our great ape ancestors, and it is probably one reason many people today suffer from what are known as the diseases of civilization: gout, diabetes, obesity, high blood pressure, and heart disease.[32]

For our great ape ancestors spreading out in Eurasia, however, this metabolic change had a big advantage. They still had to go for two to four months without fresh leaves, fruits, or nuts, depending on the region, but now they could fatten up first so they were better equipped to survive the time when food was in short supply.[33] As an additional benefit, uric acid stabilizes blood pressure. This means that species with this genetic mutation find it relatively easy to stay both mentally and physically active over extended periods without tiring too rapidly, even when they go for a long time without food. Great apes that adapted better to the conditions in the more northerly regions and were therefore more advanced in evolutionary terms ended up inhabiting vast areas in Eurasia. Around at least 12.5 million years ago, probably even earlier, they split into two subfamilies: Ponginae and Homininae.

The Ponginae, modern orangutans and their fossil relatives, settled mostly in the huge eastern area, from Turkey to as far east as China. As they are pongines, not hominines, they play no further role in the subsequent path toward the evolution of humans—even if a few researchers in the past saw things differently.

The Homininae, or hominines,[34] include humans and the great apes that live in Africa today (chimpanzees, bonobos, and gorillas), together with their ancestors. These ancestors lived in the western part of Eurasia. The various species of *Dryopithecus*, as well as *Anoiapithecus*, *Pierolapithecus*, and other European great apes, also belong to this group.

During this middle phase of great ape evolution, Europe was settled mostly by the hominines. At this time the body shape of this subfamily was changing, evolving into one that would become characteristic of all more highly evolved great apes. The apes' arms grew longer, and they were able to straighten their elbows. Their rib cages widened, and their shoulder blades moved to their backs. When they balanced on branches, they no longer held their rear ends horizontal all the time but sometimes oriented them vertically.

Europe as a Gigantic Great Ape Laboratory

These anatomical changes meant that the hominines began to move differently. Instead of spending all their time walking and climbing on all fours, as the apes in Africa had done (*Ekembo*, for example), the more highly evolved species began to swing from branch to branch. They went from being nimble scutters to strong swingers. Their wrists became more flexible, and their more powerful arms made it possible for them to climb quickly and with greater ease as they scaled the trunks of trees. The anatomy of their hands and feet also adapted. Their hands were now slightly curved so that they could grip branches better and save energy when they hung from them—even when they were hanging by just one arm. Their ankles also became looser and flexed easily—ideal for an animal that wanted to put its feet on a trunk to gain traction. Their big toe was so long that it could be stretched far to the side so that these animals could grip even thicker branches with their feet.

These features meant that the apes that had left Africa for Europe could now climb up into the small branches of the trees, where they were rewarded with nutritious fruit that earlier they had mostly been unable to reach because it often grew at the ends of thin branches. As the apes' arms were longer than their legs, even when they were walking on the ground on all fours, their spines were angled slightly upward. If you watch orangutans walk on their fists or chimpanzees and gorillas walk on their knuckles, you will notice the palms of their hands do not touch the ground. Even if they wanted to walk that way, their curved fingers would not allow them to.

When you think how long evolution generally takes, these are very substantial changes in a relatively short time. Such adaptations usually happen when life-forms have to adapt to quickly changing environmental conditions and occupy a new niche—which was exactly what happened with the great apes in Europe between 14 million and 7 million years ago.

Globally, the climate was somewhat cooler and drier, but there were short-lived regional exceptions to this. Although large parts of northern Africa were increasingly becoming savannah or desert landscapes, the situation in Europe was much more complicated. At times, rainfall was only half what it is today. Large dry areas developed where few trees grew. In Spain, for instance, desert-like areas became widespread. In between, however, there were two warm and damp interludes, one from about 11 million to 9.7 million years ago and another from 9.3 million to 8.7 million years ago. Sometimes the woodlands retreated to be replaced by open, savannah landscapes, then thick woodlands reclaimed the land. Plants and animals had to adapt relatively quickly to these variations in climate.

In the warm and damp interludes, there was a lot of rain, about three times as much as we have today. Temperatures were mostly at subtropical to tropical levels with an annual average of over 68 degrees Fahrenheit (20 degrees Centigrade). The climate was muggy, more humid than anything in equatorial regions today. It must have been like being in an old-fashioned wash house. The reason was probably the newly created Isthmus of Panama, which now formed a land bridge between North and South America.[35] Tectonic forces inside the Earth had closed the gap between these two continents. Warm surface waters from the South Atlantic could no longer flow into the Pacific, and they turned and traveled northeast instead. For the first time in Earth's history, the Gulf Stream started carrying warm water to the North Atlantic. The heat released by evaporation and the substantial differences in atmospheric pressure between Iceland and the Azores led to heavy westerly winds and frequent intense rains in Europe.

The enormous quantities of rain in Europe during this wash-house phase led to the formation of large river systems, such as the Rhine and the Danube, and extensive lakes. And as the continental plates of Africa and Europe collided, mountain ranges such as the Alps, the Pyrenees, and the Carpathians were thrust upward and rose to towering heights. With their varying altitudes, they created a large number of habitats that had not been seen before, and plants and animals had to find new evolutionary responses to survive there.

It is hard to say whether it was the new diversity in the environment, the increase in the amount of water available, or their more effective genetic adaptations that allowed great apes in Europe to survive the change especially well during the middle phase of their evolutionary history and to continue to evolve remarkably quickly. What is certain, however, is that so far there have been no fossil finds of any more highly evolved apes from Africa from that time, even though many experts have obsessively searched for them.

David Begun, who in recent years has researched the evolution of great apes more thoroughly than anyone else, thinks it is unlikely that over the course of the Miocene, the time period from about 23 million to 5 million years ago, similar species of apes inhabited both Africa and Eurasia. "Finally it occurred to me," Begun wrote in his book *The Real Planet of the Apes,* which summarizes his decades of research, "that the ancestor of the African apes and humans may actually have evolved in Europe instead of Africa."[36] Europe 14 million to 7 million years ago must have been like a giant laboratory where great apes made huge evolutionary leaps forward. Then, as climate conditions became more challenging in Europe and more favorable once again in Africa, these more highly evolved great apes would have returned to Africa.

Although much of what happened at this time remains shrouded in mystery, Begun's hypothesis is very likely right. A series of spectacular finds that my colleagues and I were lucky enough to make recently in Germany shed more light on the matter. Before we move on to the third and last phase of great ape evolution, I would like to tell the story of our finds.

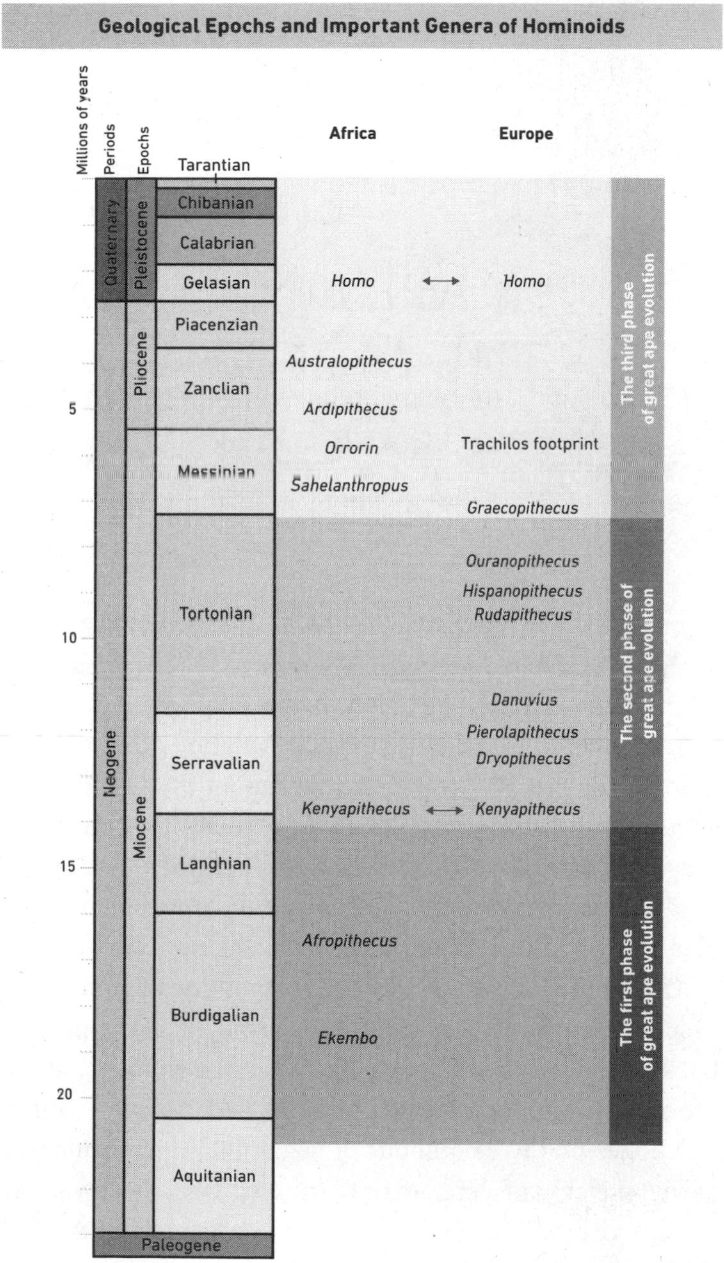
Geological Epochs and Important Genera of Hominoids
Millions of years
Periods
Epochs
Tarantian
Africa
Europe
Quaternary
Pleistocene
Chibanian
Calabrian
Gelasian
Homo
Homo
Piacenzian
Pliocene
Zanclian
Australopithecus
5
Ardipithecus
Orrorin
Trachilos footprint
Messinian
Sahelanthropus
Graecopithecus
The third phase of great ape evolution
Ouranopithecus
Hispanopithecus
Rudapithecus
Tortonian
10
Danuvius
Pierolapithecus
Dryopithecus
Neogene
Serravalian
Kenyapithecus
Kenyapithecus
The second phase of great ape evolution
Miocene
Langhian
15
Afropithecus
Burdigalian
Ekembo
The first phase of great ape evolution
20
Aquitanian
Paleogene

- 9 -

APES IN
THE ALLGÄU

Was Udo a Missing Link?

N OT FAR FROM Neuschwanstein Castle, near Kaufbeuren in the Allgäu, stands a picturesque baroque
monastery called Irsee Abbey. The ground around
the abbey is full of clay-rich rock formations that contain
seams of lignite. People have known about this lignite since
ancient times, and local residents used to dig it up, mostly so
they could use the soft brown coal for fuel when they fell on
hard times. Until recently, however, almost no one had any
idea of the other riches the clay contained.

One of the few who knew it hid something of great value
to science was the amateur archeologist Sigulf Guggenmos.[37]
Back in the early 1970s, in a clay pit in the municipality of
Pforzen a few miles from the abbey, he had discovered mammal bones that were millions of years old. They included a
partial skeleton of a precursor to the elephant, the lower jawbone of a hyena, and the remains of a forest antelope. Around

the same time, scientists from the University of Munich were also digging in the area for small mammal fossils, in an exercise completely unrelated to Guggenmos's. They, too, made discoveries, including many unknown species and genera of rodents and insect-eaters. They published an article in 1975 describing a lost world rich in species diversity.[38] In the article, they also described the rock and noted that a distinguishing feature of the clay layer was that it was full of small lumps of black coal—an observation that turned out to be a mistake of monumental proportions.

For at least thirty years, no one other than Guggenmos was interested in the paleontological treasures hiding underground near Kaufbeuren. Instead, hundreds of tons of clay were dug out every year from a pit known locally as Hammerschmiede ("Hammer Mill") and sent to a factory where the clay, along with everything it contained, was fired to make bricks. When I first visited the pit in 2006, I quickly found the distinctive layer rich in black material. It formed an undulating strip about 16 feet (5 meters) wide and about 16 feet (5 meters) up the wall of the pit. The small lumps of what was allegedly coal—some a little longer than my hand is wide—were soft and brittle. They crumbled when I pressed them between my fingers. Lignite, which is mostly firm to the touch, does not do that. Intrigued, I got out my magnifying glass. The image that greeted me under ten times magnification made me even surer of what I had suspected all along. I was looking at open-grained, spongy structures: blackened bones.

A careful initial examination revealed that the odd-looking clay layer was the buried channel of a water course millions of years old, and it contained innumerable fossil remains of both small and large animals. For experienced paleontologists, that was not in itself anything particularly remarkable, and starting in 2011, we began to unearth fossils every year in increasingly extensive digs. Finally, during a dig in the summer of 2015, our persistence was rewarded with a sensational find: there in soft gray clay lay the tooth of a great ape.

That moment changed everyone's view of the significance of Hammerschmiede. All the scientists involved immediately realized that we were digging at a site that could contribute important information about great ape evolution.

We had an arrangement with the pit operator. For a few weeks a year, we were allowed to carry out our dig in the area around the channel, and we agreed to stay out of the way of the pit workers as best we could. Suddenly, for the first time since we had started digging, the mining operations, which swallowed up a considerable quantity of the scientifically valuable seam every year, seemed like a serious threat to our work. Priceless ape fossils could disappear forever. And they were crucial clues to understanding the origins of humanity. But there were limits, both personal and professional, as to how much and how fast we could dig. The *Deutsche Forschungsgemainschaft* (the German Research Foundation), the most important funding agency for scientific research in Germany, denied my application for funds to co-finance a salvage dig because, the committee explained, they finance science and not excavation work.

Despite this setback, the following year my team and I decided on an unusual course of action to save what we could in this dire situation: we would dig up as much of the channel sediments threatened by the mining operations as possible.

Udo Lindenberg and Our Sensational Find

It was the morning of May 17, 2016, a sunny day in spring, and I was driving from Tübingen to the Allgäu in a Volkswagen van with my doctoral student Jochen Fuss. All the radio stations we tuned in to were covering the same story: Udo Lindenberg, Germany's most famous rock musician, was celebrating his seventieth birthday today. We arrived at Hammerschmiede in a good mood, Lindenberg's hit songs still running through our heads. An excavator operator was waiting for us. Our plan was for the excavator to remove as much of the channel sediment as possible, and we would store it safely in a nearby facility. We could then go through the sediment to look for fossils at our leisure. The operator cautiously moved the big shovel on his excavator forward, almost as though he were feeling his way, trying not to disturb any larger bones that might be hidden in the clay. Despite the care he was taking, digging with an excavator was going to do a lot of damage, and we knew it, but I could not see any other way to save at least a few important fossils from being baked into bricks.

A little while later, the excavator had deposited 25 tons of sediment into what we thought of as our treasure chamber. The pile looked disappointingly small, and I was sorry that I could not save any more of the formation. But the storage facility was completely full and another one was not available. When we looked at where the channel had been dug out, it seemed as though we had barely scratched the surface. Time had run out, and the next day the excavator would be back to carrying channel sediment to the furnaces where the bricks were made. We decided to spend what we had left of the day digging around the sediment a bit by hand. We got our pickaxes and set to work in the spot where the excavator had left its mark.

I hit the rock hard three times with my trusty pickax to loosen it and then turned the pieces over carefully. What I saw next took my breath away. Protruding out of the light-gray clay was a piece of dark brown bone. The two sturdy teeth embedded in it reflected the sunlight. The size and shape of the teeth left me in no doubt: this was the lower-left jawbone of a great ape. Fuss hurried over, and we looked at each other. Both of us were grinning from ear to ear. We had been systematically digging in Hammerschmiede for five years now. We had carefully sifted through the channel sediment inch by inch using spatulas, needles, and paintbrushes to make sure we did not overlook anything. And then, when we had no choice but to delve in with pickaxes and an excavator, we unearthed something really sensational.

After we had cleaned and packed our priceless find, we continued searching, fueled by our excitement. We diligently turned over every fragment of rock the excavator had left behind. And we were rewarded with another lucky find. In one of the fragments, we found yet another ape tooth. It is moments like these that really get a paleontologist's heart racing.

The tooth turned out to be the second molar from a lower-right jawbone. The single tooth we had found the year before was a lower-right wisdom tooth. Might the lower-left jawbone and the two single teeth from the right jawbone possibly belong to the same individual? If that were indeed the case, we had a once-in-a-lifetime chance of digging up the scattered pieces of a single ape skeleton. On our return journey, we talked excitedly about the likelihood of that happening. The car radio was still covering the same story: Udo Lindenberg. He was turning out to be the patron saint watching over the most significant paleontological discovery made in Germany for decades. We quickly decided to give our ape—who was in fact a little male as we later discovered—the nickname Udo.

From that day on it was clear we had to double down on our efforts and to dig much faster and cover a much wider area. There was nothing we could do to bring back the many hundreds of tons of clay we had lost along with all the historical details they contained. Any spoonful of sediment could have held a puzzle piece that might have belonged to Udo's skeleton. And now we had to do everything in our power not to lose even more material without examining it first. But how could we manage that? Dig as fast as the clay was being mined? That would require many scientists and workers to be on-site for many months of the year. We had a handful of students and no extra money at our disposal, so that was not going to happen.

Because paleontological finds do not enjoy protected status in the German state of Bavaria, we could not look to the state for assistance. Ironically enough, it all hinged on where Udo fit in evolutionary terms, because finds of archeological significance are extended some protection by the state, and the boundary between paleontology and archeology is marked by the advent of civilization. If Udo used tools—as we know chimpanzees do—then the finds at Hammerschmiede would be considered part of Bavaria's archeological heritage. However, to learn more about Udo so that we could make the appropriate arguments, we would need more material. We looked at our situation from every possible angle, but there was no way around it—we had to find a different approach.

I had a brainwave. We would start a citizen dig. These initiatives have been proven to attract people of all ages from a range of different educational backgrounds, all keen to connect with their inner scientist. I started looking for members of the public who might be interested in helping. I used internet forums and a network I was already plugged into. I offered people the chance to take part in a scientific adventure for free. On some archeological projects, hobby researchers and tourist treasure hunters pay a lot of money for an opportunity like this. But the only things the excavators had to bring with them to this dig were interest and curiosity— and copious quantities of patience.

Luckily, lots of people signed up. More than fifty volunteers participated in three-month-long digs in 2017 and 2018. The amateur paleontologists ranged from nine years old to seventy-five years old. Young people and families with children took part alongside retired couples. There were friends and parents of colleagues, people from other parts of the country who had learned of the dig through word of mouth, and, of course, students from Tübingen and elsewhere. Our strategy of contacting geology students at many different German universities through social media had been particularly successful.

In those two years, we analyzed more than 7,000 cubic feet (200 cubic meters) of sediment and made over five thousand confirmed finds. We verified more than one hundred different vertebrates, including many species that had never been seen before—a treasure trove of fossils of the kind rarely found anywhere in the world. Thanks to these finds, we could sketch the rough outlines of a long-lost tropical ecosystem with fish, giant salamanders, turtles, snakes, and birds that had died out long ago, along with many mammals from mice to rhinoceroses and elephants. And to our delight, we also made thirty-six finds relating to apes, all of which were unusually well preserved.

Generally, the most common great ape remains scientists find are teeth. Bones are often disturbed by scavenging hyenas, by weathering, or as a result of unsophisticated excavation techniques, which means that there are frequently fragments but seldom complete bones. In Hammerschmiede, however, we found complete bones from almost every part of the body. Most, that is to say, twenty-one of them, could be traced to a single individual: our Udo.

We had at least 15 percent of his skeleton: hand and foot bones; bones from his spine, legs, and arms; and even his knee cap. And Udo wasn't the only individual we unearthed. Fifteen finds could be attributed to three more great apes: two females, one larger and one smaller, and a juvenile. Might we be dealing here with the oldest original family ever found?

IF WE COULD get into a time machine and travel back to Udo's world and immerse ourselves in the Allgäu as it was about 11.62 million years before our time,[39] we might be greeted by the following scene: It is an incredibly hot fall morning. The outline of the Alps is clearly visible on the horizon about 30 miles (50 kilometers) to the south. The highest peaks, those above 8,000 feet (2,500 meters), are dusted with the first snows of the season. The air shimmers in places over open grasslands that stretch to the foot of the mountains, broken only by the occasional tree. Now and then, a hot gentle breeze brushes the stalks of the dry brown grasses. The vegetation is thirstily awaiting the first rains of winter. The only food still available for the numerous herbivores that live here are shrubby plants and bushes. You can make out small herds of Munich wood antelopes.[40] From a distance, these slender creatures with their two powerful pointed horns look like well-armed deer.[41] A female gomphothere uses her trunks and tusks[42] to scare a group of muntjaks away from her young one. The tiny swift-footed deer flee in a wild zigzag pattern. Their short forelegs and the acute backward angle of their forked antlers make them look awkward and panicked.[43]

The higher the sun climbs, the fewer animals there are to be seen in the hot savannah landscape. The only ones that show themselves, apparently unfazed by the heat, are narrow-footed rhinoceroses[44] and pheasants[45] checking for seeds brought to the surface after the rhinoceroses have churned up the ground. A few ponds overgrown with cattails[46] contain the pitiful remains of moisture from powerful, long-lasting summer thunderstorms.[47] Numerous snapping turtles hide in the cattails,[48] lunging at almost anything that crosses their paths: snails, frogs, and even careless waterfowl. When there is no longer anything to eat, they will move on to the next pond. Not far from one of the ponds, a lion-sized bear-dog[49] is pulling apart a wild pig it has caught and cracking open its bones.[50] It's best to give this powerful predator a wide berth. It's likely the most dangerous one we're going to meet here. The bear-dog is on high alert and aggressive, because at any time a troop of hyenas[51] could steal its prize.

A few miles on, a dark green, densely vegetated artery winds its way through the landscape. The meandering strip of lush growth exists thanks to a narrow water course that originates in the foothills of the Alps.[52] In the distance, we can hear the mating calls of "hooked beasts."[53] In the soft ground, we can make out the tracks of single-hoofed animals; in this case, the tracks are from large forest horses.[54] In a stand of beeches,[55] a magnificent panda bear[56] is sleeping in the fork of a branch, while a large flying squirrel[57] that looks like a magic carpet glides playfully from branch to branch. We're working our way through thick undergrowth and arrive at the river. At this time of year, it's barely 16 feet (5 meters) wide. The water stands rather than flows. At its deepest point, it barely reaches our knees.[58] It's alive with fish. A giant salamander, 5 feet (1.5 meters) long,[59] has no trouble snapping up its favorite meal—a catfish[60] the size of your palm squeezed tightly into a cool muddy hollow.

Suddenly there's a deafening noise. It sounds like a human war cry! We stop. Then, as the screaming continues, we start to run in the direction of the alarming cries. If we synchronize our bursts of speed with its screams, whatever it is will not hear us coming.

When we are close enough to the action, a dramatic scene unfolds before our eyes. On a thick, angled branch some distance above the ground sits a cat ready to pounce. It's the size of a leopard,[61] and it is baring its formidable canines. Directly beneath, in a tangled curtain of lianas, we spot the source of the screaming: a male great ape. Perhaps it is Udo himself.[62]

To make himself look more intimidating, the ape is hanging onto the vines with his hands and feet, his body vertical. He is sticking out his chest in a menacing fashion. He's stretched out to his full height, about 3 feet (1 meter), to make himself look more imposing. His knees and hips straight, he

seems to be standing upright in the vines—an unusual posture for an ape. With short, loud calls, he's clearly trying to
send a message to the predatory cat. You don't stand a chance
in the lianas. This is our domain. The cat has its sights set not
just on the small male but on the whole group. Females and
young are "standing" a short distance below him in the network of vines. The cat is swift and agile on the tree branches,
but its paws and claws do not function as well in the unsteady,
rope-like lianas. After a good hour of laying siege to the group,
it gives up and slinks away. It takes a while for the ape family
to calm down. And now they surprise us once again. The four
of them proceed in single file along the sturdy branch where
the cat was recently sitting. Walking on two legs, their knees
and hips straight, they look almost like people as they head
toward the trunk. Their long, flexible big toes have a powerful hold on the branch to stabilize their upright bodies. Then
they use their long arms to lift themselves up into the next
level of branches. We've never seen great apes move that way.
How unusual!

The Missing Link Between
Great Apes and Humans?

Let's leave our thought experiment and the far distant past
and return to the present. Time travel helps us make connections and come up with hypotheses, but facts are what count
in science. With all the euphoria about our unusual ape finds,
we had to answer the following questions: Do they belong to
one and the same species? Are they the same as fossils that
have already been found?

It soon became clear that the bones did not match any known species. The spectacular finds from the Allgäu were the remains of a species and genus of ape that had never been described before. My colleagues and I tried to come up with a suitable scientific name. We didn't have to think for long. We decided on *Danuvius guggenmosi*. Danuvius is the name of a Celtic river god. More than two thousand years ago, Celtic tribes lived in the area now called the Allgäu. And the names of the Danube and Don Rivers come from Danuvius. The species name *guggenmosi* honors the recently deceased discoverer of the first fossils from Hammerschmiede, Sigulf Guggenmos.

Although *Danuvius guggenmosi* is only 300,000 to 200,000 years younger than the three species of great ape found around Barcelona (*Dryopithecus, Anoiapithecus*, and *Pierolapithecus*), it is very different from them.

The extremely sturdy cheekbones are high above the jaw and cover extensive sinus cavities. The arched palate and many features of the teeth also show that *Danuvius* belonged to a different and more evolved group of great apes than those found in Catalonia. With these skull features, *Danuvius* had more in common with significantly younger European great apes,[63] and especially with the great apes found in Africa today: gorillas, chimpanzees, and bonobos. What is unique to *Danuvius* is that the lower part of the face is truncated and does not stick out very far. This is a feature that more closely resembles the faces of early hominins than of great apes.

It must be said that Udo, our male *Danuvius*, is much smaller than any of his living relatives in Africa. He is only about 3 feet (1 meter) tall and weighs about 66 pounds

(30 kilograms).[64] Those are about the same values calculated for the famous Lucy. The two female *Danuvius* individuals, at about 42 and 37.5 pounds (19 and 17 kilograms) respectively, are considerably lighter.

Using Udo, we did an in-depth study of the physical proportions of the genus *Danuvius*. As with all living great apes with the exception of humans, his arms are longer than his legs. The relationship between his lower arm and lower leg is similar to that of chimpanzees and bonobos, whose forelimbs are 10 percent to 20 percent longer than their hind limbs. The arms of gorillas, and especially of orangutans, in contrast, are considerably longer than their legs.[65]

Danuvius has a couple of features that are particularly striking: large powerful thumbs and big toes. The bone of Udo's big toe is spread wide to one side so it can be flexed against the sole of its foot. The powerful grip of our hand works in much the same way when, for example, we want to pull ourselves up using a handrail. Udo, therefore, could grip things firmly not just with his hands but also with his feet. In relation to his body weight, his big toe is longer than that of a chimpanzee, a bonobo, or even a human—even though we humans have the longest big toes of any of the living great apes. The sideways orientation of his big toe bone is typical for all tree-dwelling monkeys and apes.[66] With the Allgäuer ape, however, this sideways rotation is more marked than in any other living or fossil ape species.

When Udo balled his foot like a fist, he could probably generate enormous force. Thanks to the flexibility of the last joint in his big toe, he could grip even very small or thin objects firmly with his feet. That made him perfectly equipped for life in the tangle of lianas and smaller branches. And that is probably why the spaces between trees and branches where mostly vines grew was *Danuvius*'s preferred habitat. Those spaces were where this ape could find reliable protection from tree-climbing predators.

What was particularly unusual was that Udo's partial skeleton included two vertebrae from the portion of the spine where his ribs were attached to form his chest—the topmost vertebra at the height of the collarbone and one from farther down. The anatomy of the upper vertebra revealed that *Danuvius*'s rib cage had already expanded. The lower chest vertebra was wider and appeared to already be functioning as a lumbar vertebra. These observations are hugely significant, because they prove that *Danuvius*'s lower back, in contrast to all living great apes' backs, was elongated. It was the elongated lower back of early hominins that allowed for an S-curve in the spine, without which it is impossible to keep one's balance when walking upright. The anatomy of *Danuvius*'s spine was consistent with features of its thighbone and shinbone that prove *Danuvius* bore its weight on hip and knee joints that could be extended and straightened.

And there we have it. The most unusual features of this new species were its ability to stand straight with its knees and hips fully extended and a lower back that curved gently to lend it stability. When today's great apes stand on two feet, they keep their knees and hips bent.[67] Their lower backs are inflexible and too short for them to be able to stand upright with extended hips. When they climb, their hips and knees bend, as do their ankles, and their long arms do most of the work. The bones we found from *Danuvius* showed its body was built completely differently from the bodies of chimpanzees, gorillas, and orangutans. When Udo climbed, his legs did most of the work, and he used his arms and hands to hold on. These features make *Danuvius* one of the candidates to be the last common ancestor of both great apes and humans. To put it somewhat flippantly, from the waist up he looked like an ape and from the waist down he looked like an early hominin.

We do not yet know whether Udo and others of his species spent time on the ground or walked from tree to tree. To figure that out, we would need more bones from his feet. Perhaps they simply climbed through hanging vegetation. That way, with their ability to hold onto things and use the lianas as climbing aids, it was easier for them to reach fruit hanging at the outermost ends of branches.[68]

Danuvius certainly could also use its arms to swing from branch to branch. However, as its comparatively straight finger bones and its elbow joint show, that was not its specialty. *Danuvius* likely spent little time hanging from the "ropes" and more time standing on a dense mesh of lianas and branches. This behavior is an adaptation that has never before been recorded in great apes. We assume that this was the prototype for a method of locomotion that led to the permanently upright gait of humans, because the structural design of chimpanzees and gorillas is much too well adapted for locomotion that depends on the arms for swinging from tree to tree and climbing to be the jumping-off point for our bipedalism.

It is this unique form of locomotion that points to *Danuvius* as a primordial ancestor of the last precursor shared by both chimpanzees and humans and a genuine missing link between great apes that walk on four legs and humans that walk on two. The German ape from the Allgäu is therefore a key find if we are to understand how humans—in the words of that book from my childhood—"rose above the animal realm."

AROUND 7.4 MILLION years ago, barely 5 million years after the time of *Danuvius,* increasing ice cover at the poles ushered in the third phase of great ape evolution. The western Antarctic was by this time completely glaciated and the whole of the continent was covered with ice. Greenland was also covered with a thick layer of ice for the first time. With the complete glaciation of the polar regions, the world as we know it today was formed.

This is the world of *Graecopithecus freybergi,* whose 7.2-million-year-old remains were found in Pyrgos in Greece and Azmaka in Bulgaria.

Part 3

THE CRADLE OF HUMANITY: AFRICA OR EUROPE?

– 10 –

THE PRIMAL ANCESTOR

Still an Ape or an Early Hominin?

IN AN ARTICLE published in 2017, I, along with my colleagues Nikolai Spassov, David Begun, and Jochen Fuss, put forward the hypothesis that *Graecopithecus* was no longer a great ape but the earliest potential hominin.[1] This hypothesis was based on the fact that this species had typical hominin teeth. Apart from upright gait, this is one of the few characteristic features of the human lineage (Hominini) on which experts more or less agree.

The idea was well received by the general public but, as expected, got a mixed reaction in the world of academia. The places where El Graeco was found (Greece and Bulgaria) did not fit the accepted idea that the crucial advances in human evolution had taken place in Africa and nowhere else.

It is also generally thought that the geographic distribution of a species cannot confirm a group's lineage and therefore can be of no help in answering the question of whether El Graeco was an early hominin or not. The only useful evidence when tracing a group's origins are physical features or what lies in its genes. It is impossible, however, to extract genetic material from fossils that are millions of years old. And because we have not found any footprints from *Graecopithecus* to give us more information about the way it walked, the only way we can answer the question of whether El Graeco really was the oldest known hominin is by using physical evidence from fossils. It is clear that we need more finds if we are to pin down El Graeco's position in the evolutionary tree of life more precisely.

The Danger of Misinterpretation

The central question in our discussion over El Graeco's evolutionary position is this: Is *Graecopithecus* more closely related to chimpanzees or to humans—that is to say, is it still a great ape or is it sufficiently evolved to be considered an early hominin? It is not easy to find a satisfactory answer to this question. There are three challenges.

The first challenge is what scientists call homoplasy, which refers to a feature that occurs independently over the course of evolution twice or even multiple times. A couple of examples will be helpful here. The trunk is a feature shared by elephants and tapirs even though these animals are not closely related. Their trunks are features that developed independently from each other. Fins, to give another example, are adaptations for life in water that arose in fish, ichthyosaurs, whales, and many other vertebrates that live in water but have no common evolutionary origin. These two examples are obvious, but with many anatomical features it is extremely difficult to recognize the parallel evolution of specific attributes, and experience has shown that homoplasy lurks everywhere, inviting mistakes in interpretation when scientists try to trace the lineage of primates.

The second challenge is caused by the close similarities between two evolutionary lines when they begin to split. Today, a wide variety of different features makes it easy to distinguish between humans and chimpanzees. However, when their last common ancestors split into two isolated populations over 7 million years ago, the external appearances of the members of both groups were still exactly the same. It was only after a geographic separation over a long period of time that the two populations developed increasingly diverse features based on random changes in their genetic material and contrasting living conditions. And only then can they be clearly distinguished from each other. But even many millions of years after a split, the members of both lines were anatomically still very similar and could possibly still interbreed. The huge differences between chimpanzees and humans today are the result of at least 7 million years of independent chimpanzee evolution plus 7 million years of independent human evolution.

The third challenge is the notoriously incomplete "fossil record." This is what paleontologists call the sum of all existing, scientifically documented fossils sorted by where they fit chronologically, geologically, and geographically. For most mammal fossils, all we have are teeth. Only for a very few great ape fossils do we also have a skull, pelvic bones, or vertebrae, and we never find a complete spine. It is the lot of the researcher that important pieces of the puzzle are almost always missing from a dig. Also, information about soft tissue—fossilized fat layers, for instance—is rarely available. Paleontologists therefore have to be masters at completing puzzles that are missing most of their pieces.

As far as El Graeco is concerned, we are left with the following uncertainties:

Graecopithecus could have evolved features that are typical of humans independently, without being part of the human lineage. That is highly improbable but still possible, and that is why we have called him a "potential early hominin."

If *Graecopithecus* was an early hominin, he was only slightly different from the original ancestors of chimpanzees and, so far, we do not know what these looked like.

As we have described only two *Graecopithecus* fossils so far, a lower jawbone and an upper premolar, the picture we can sketch of this species is, of necessity, rather vague.

Did El Graeco Walk on Two Feet?

Probably the most important feature that unifies all the members of the human lineage and that, as far as we know, did not arise more than once is upright gait. Walking on two legs, also called bipedalism, is the revolutionary development that marks the beginning of human evolution. Proof of bipedalism is absolutely necessary if a fossil is to be confirmed as a hominin. The most impressive proof comes in the form of fossil footprints. These, however, are extremely rare. For the majority of fossils of early hominin species, anatomical details of the feet and legs provide valuable clues. The changes that accompanied bipedalism affected a wide range of body parts involved in locomotion: bones, musculature, sinews, and other physical features that affect biomechanics.

Even if a few monkeys and most apes can support themselves on two feet for a short while,[2] only humans are capable of doing so for an extended period of time. Indeed, for long distances, we cannot do anything but walk on two feet. Whereas apes have to make a big effort to walk upright for just a few feet, we find it extremely difficult to walk on all fours.

The most important anatomical criteria for upright gait can be summarized as follows: Members of the human evolutionary line carry their body weight on two legs only. As the arms no longer play any role in bearing weight during locomotion, they are significantly shorter than the legs. The longer legs are the result of an elongation of the shinbones in particular. The head is balanced directly above the neck and is no longer supported by powerful neck muscles. The hole where the spinal cord enters the skull is therefore located underneath the skull instead of at the back of it. As the arms no longer restrict the chest cavity, which happens when apes walk on all fours, the human rib cage is broader than that of an ape. The shoulder blades move up and away from the sides of the body and are now located completely on the back. To cushion vertical impact, the spine is S-shaped. The pelvic girdle becomes shorter and wider, and the two broad bones at the top of the pelvis form a bowl shape. That shortens the distance between the sacrum, the triangular bone at the bottom of the spine, and the hip joint, which lends more stability to the whole hip area. The musculature of the rear end bulks up to allow bipeds to straighten their hips and stand upright. The leg muscles bulk up as well. Heavier muscles, together with longer, and therefore heavier, leg bones lower the body's center of gravity. For a more secure stance, the thighbones angle slightly inward so the knees end up directly beneath the body's center of gravity.

Upright Gait

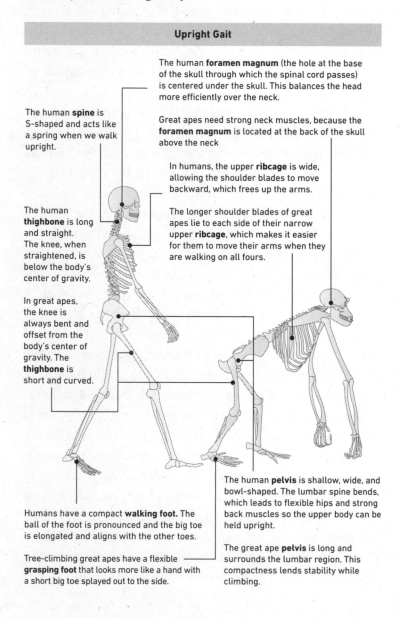

The human **foramen magnum** (the hole at the base of the skull through which the spinal cord passes) is centered under the skull. This balances the head more efficiently over the neck.

Great apes need strong neck muscles, because the **foramen magnum** is located at the back of the skull above the neck

The human **spine** is S-shaped and acts like a spring when we walk upright.

In humans, the upper **ribcage** is wide, allowing the shoulder blades to move backward, which frees up the arms.

The human **thighbone** is long and straight. The knee, when straightened, is below the body's center of gravity.

The longer shoulder blades of great apes lie to each side of their narrow upper **ribcage**, which makes it easier for them to move their arms when they are walking on all fours.

In great apes, the knee is always bent and offset from the body's center of gravity. The **thighbone** is short and curved.

Humans have a compact **walking foot.** The ball of the foot is pronounced and the big toe is elongated and aligns with the other toes.

Tree-climbing great apes have a flexible **grasping foot** that looks more like a hand with a short big toe splayed out to the side.

The human **pelvis** is shallow, wide, and bowl-shaped. The lumbar spine bends, which leads to flexible hips and strong back muscles so the upper body can be held upright.

The great ape **pelvis** is long and surrounds the lumbar region. This compactness lends stability while climbing.

Comparison of a Human and Chimpanzee Foot

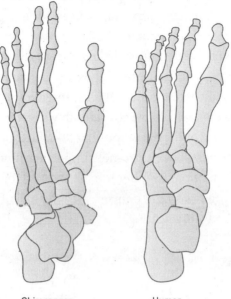

Chimpanzee   Human

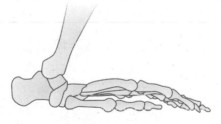

Chimpanzee

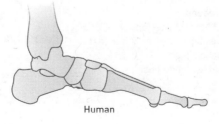

Human

There were also profound evolutionary changes in the anatomy of the feet as great apes evolved into humans. Hominin feet are no longer appendages used for grasping things. They now facilitate a stable stance on two legs and, even more important, efficient, swift, balanced, forward locomotion. In contrast to the big toe of apes, the human big toe is no longer splayed out to the side. It is oriented forward, parallel to the other toes, which are considerably shorter in comparison. At the base of the enormous big toe, the foot developed a ball, which has a new, extremely important function in the sequence of motion. The big toe, as the last toe in contact with the ground, works in concert with the ball of the foot to propel the body forward.

Changes in the Shape of the Foot From Great Ape to Human

Footprint of a modern human

Footprint of a great ape

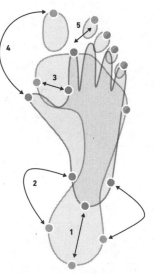

1 The midfoot lengthens.
2 The heel widens.
3 The ball of the foot develops.
4 The big toe aligns with the other toes and increases in size.
5 The second toe lengthens.

In later stages of evolution, early humans such as *Homo erectus* developed an arch to the foot to absorb the shock of repeated, forceful contact with the ground. That was mostly necessary because the lifestyles of early humans meant the ability to run long distances became much more important than it had been for early hominins (see Chapter 20).

Even though the list of features required for upright gait is long, there are significant challenges to recognizing the early stages of bipedalism. Even with well-documented skeletons, such as that of Lucy, there was heated scientific debate.[3] The controversy mostly centered around whether this species of early hominin could be said to have developed a modern style of bipedalism or whether Lucy still spent most of her time living in trees.

In many finds that mark the beginning of the human ancestral line, crucial anatomical regions are missing. In others, a few characteristic features have been found, such as the fragment of a shinbone (*Australopithecus anamensis*), a section of thighbone (*Orrorin tugenensis*), or a small foot bone (*Ardipithecus kadabba*); however, because of the complex interaction of many regions of the body in upright gait, such sparse evidence is often not sufficient for us to be able to say definitively whether these species spent most of their time walking on two legs or not.

As mentioned, the surest evidence of upright gait is a fossil footprint. But such a discovery, one of the most priceless historical records a paleontologist could find, would be an exceedingly rare stroke of luck. Until recently, the fossil footprints of early hominins had been found only once: 3.6-million-year-old footprints from Laetoli in Tanzania that definitively ended the debate about Lucy's upright gait. Today, more mysterious tracks of a biped, dating back millions of years, have emerged—this time on the island of Crete.

– 11 –

FOSSIL FOOTPRINTS
FROM CRETE

Puzzling Prints of an Ancient Biped

I N 2002, THE Polish paleontologist Gerard Gierliński
was vacationing with his girlfriend near Trachilos, on the
northwest coast of Crete. One day, when they were walking not far from the water's edge, Gierliński noticed a flat slab
of rock slightly angled toward the water with unusual elongated depressions on its surface. It was difficult to decipher
the marks in the glistening light. Vacationers had probably sat
on the rock numerous times to look at the sea without even
noticing them. Gierliński, however, was familiar with dinosaur tracks and realized that these must be fossil footprints,
although clearly not from any kind of extinct giant lizard. The
discovery piqued his curiosity. He recorded the GPS coordinates for the location of his find, took a few photos, and
decided to come back one day to take a closer look.

In 2010, Gierliński finally told his colleague Grzegorz
Niedźweidzki about the prints, and they developed the

hypothesis that they could have come from a primate that walked upright, perhaps even from an early hominin. An international research team was formed to examine the find in more detail.

Per Ahlberg, a renowned Swedish expert on vertebrate fossil footprints, took the lead. At the site, the researchers scanned the slab of rock using a laser accurate to five-hundredths of a millimeter so they could create three-dimensional images of the prints. In an area of only 32 square feet (3 square meters), they eventually found fifty of these depressions, twenty-eight of which they could reliably identify as footprints. The impressions were between 4 and 8.5 inches (10 and 22 centimeters) long and 1.2 to 2.75 inches (3 to 7 centimeters) wide. To capture even more details, they made silicone rubber casts of a few of the more promising-looking tracks. That way, they not only preserved the footprints but also preserved them in a form that allowed them to take further precise computerized measurements in three dimensions—a technique known as geometric morphometrics that is also used when analyzing fossils.

It quickly became apparent that this was a find of monumental proportions. The originators of the tracks must indeed have walked on two feet, because there was no evidence of tracks from forelimbs at the site. That eliminated the possibility that four-legged animals had walked across and just stood up on their hind legs for a while. Added to that, after analyzing the precise measurements, the researchers found that the prints were looking increasingly like human prints. Most of the tracks showed the clear outlines of five toes. The big toe was sturdy with a broad tip. Like a human big toe, it was aligned with the other toes and stuck out a bit farther forward than they did. The other toes were short and got smaller the farther from the big toe they were. There was also evidence of a pronounced ball to the foot. Both the ball of the foot and the big toe left deep impressions in the sand. That indicated that the transfer of weight went from the outside of the foot to the inside, and the ball of the foot and the big toe pushed off to create the forward momentum for walking. As mentioned, this is the most efficient way to walk on two legs, because the body's center of gravity stays in the middle if you push off with the big toes.

In addition, there was not even a hint of a claw mark on any of the toes. That ruled out all mammals that walk on the soles of their feet (bears, for example), except apes and monkeys, as possible originators of the prints. But these could not have left prints like these either, because all apes and monkeys, including great apes, had, and indeed still have, a shorter, narrower, and more pointed big toe that starts farther back on the foot and splays out to the side. Furthermore, monkeys and apes have no ball on the foot, their remaining toes are elongated, and their middle toe is the longest. And so there is no doubt that the tracks at Trachilos most closely resemble footprints made by humans and their extinct relatives—that is to say, footprints made by bipeds that walked upright.

The Trachilos tracks do, however, show differences from the footprints of anatomically modern humans. They are smaller and generally more compact. Most obviously, the center of the foot is shorter and the heel is relatively narrow. Also, there is no arch. You could say that whoever left these prints had flat feet. However, hominins did not fully develop arches in their feet until the first members of the genus *Homo* came on the scene about 2.5 million years ago in Africa.[4]

Where do the tracks from Crete fit in the evolutionary time line? An obvious first step to answering this question is to compare them with the famous footprints from Laetoli in Tanzania. These fossil imprints were discovered in 1978 by an excavation team working for the grande dame of paleoanthropology, Mary Leakey. Casts of the Laetoli tracks are on display in many museums today because they are regarded as the oldest proof we have for upright gait. They were made when early hominins walked through damp volcanic ash 3.66 million years ago. At the site, there are about seventy prints from a number of individuals spread over a distance of about 90 feet (27 meters). Today, most researchers consider them to have been made by *Australopithecus afarensis*, the same species of early hominin that included the famous Lucy.

The Trachilos footprints share many features with those found in Tanzania. For example, they show a sturdy big toe with a broad tip that aligns closely with the other toes, toes that get progressively smaller from the inside to the outside of the foot, and a ball that leaves a deep impression in the ground. These are all characteristics that point to the fact that whoever left these footprints walked on two legs.

Tracks on a Dried-Out Seashore

But when exactly did the unknown walker leave impressions of its feet in the Cretan sand? To find out how old the tracks were, the researchers first had to date the sandstone in which they were preserved. And here, it turns out, they were lucky with the location of the site. Over thousands of years, waves from the Mediterranean had carried away the layers of rock directly above the site, while only a few yards away some boulders blocking the waves' path still rose many feet above the sandstone layers. These boulders contain a geological archive that documents the turbulent ancient times in which the enigmatic beach walkers lived. In these boulders, in layers above the sandstone in which the tracks were imprinted, Ahlberg and his colleagues found coarser layers of sediment composed of angular rocks and other debris. Even a layperson could easily make out this clearly demarcated boundary layer.

You get coarse rock like this when base layers are eroded by moving water or displaced by landslides. The loosened material is then carried away and deposited elsewhere. Geologists call this breccia. When this layer was formed, the site where the prints were found must have been some distance from the sea and no longer on the beach. But that's not all. If you follow this coarse breccia farther into what today is dry land, you find above it a layer of a type of extremely fine-grained, homogenous rock that is deposited nowhere else but in a deep ocean basin. There is only one conclusion that can be drawn from this astonishing juxtaposition of sediment layers: the beach walker lived just before a time of immense sea-level change.

Researchers know this phase only too well, and it happened throughout the Mediterranean region. They call it the Messinian Salinity Crisis. Back then, the Mediterranean dried up almost completely, although later water from the Atlantic Ocean returned. This means the coarse breccia was formed 5.6 million to 5.3 million years ago, when the Mediterranean had already receded a long way. Eventually, it refilled like a giant bathtub. The sea level this time was higher than before, and the site where the prints were found was submerged. This was when the deepwater sediments were deposited, 5.3 million years ago.

For Ahlberg's research team, the geological evidence confirmed that the Trachilos footprints had to be more than 5.6 million years old. And there was evidence that they could have been made even earlier than that. When the researchers looked more closely, they found that the thin sandstone layer that held the prints was embedded in very fine limestone deposits made from the fossil skeletons of minute plankton-like organisms.[5] In the centuries before and after the mysterious biped passed by, the sea level must have risen slightly again, for the plankton are evidence of a shallow lagoon lying over the site where the find was made.

This type of plankton-like organism mostly disappeared from the Mediterranean 5.97 million years ago, right at the beginning of the Messinian Salinity Crisis. The reason for their disappearance was that as the salinity of the water increased, the environment for the plankton deteriorated, and eventually they died. Therefore, the Trachilos tracks more likely date back more than 6 million years.

In those times, Crete was still joined to the Peloponnese to form a long peninsula that curved around a warm, shallow basin filled by the Cretan Sea. Today, about 170 miles (275 kilometers) from Trachilos as the crow flies, you can find Athens and the site where El Graeco was found on what would have been the sea's northern shore. It is not too far-fetched to think that *Graecopithecus* might potentially be one of this unknown beach-walker's ancestors. According to the researchers, the sandstone that held the Trachilos prints could have been deposited at the mouth of a river. Perhaps the biped followed this river down to the sea. After all, ocean beaches and areas of brackish water offer the promise of plentiful food. Nutritious mussels, snails, and seaweed are there for the taking. All you have to do is bend over and scoop them up.

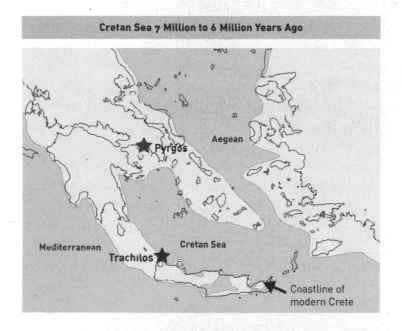

Cretan Sea 7 Million to 6 Million Years Ago

The research results documenting the footprints from Trachilos were published in July 2017, only two months after the scientific paper on *Graecopithecus freybergi*. For our research group, the results were sensational, because they strongly rattled the cage of accepted thinking that the earliest ancestors of humans lived in Africa and nowhere else. The Trachilos find is the oldest direct evidence by far of upright gait. It shows a relatively modern foot anatomy that is at least 2.3 million years older than the tracks from Laetoli. And the imprints are from human-like beings from Europe, not Africa!

It is true that whoever or whatever left the footprint in Greece had no arch in its foot, and some of the tracks from Laetoli show clear signs of an arch. However, Lucy—unlike others in her species, the ones that left the footprints in the ground at Laetoli—also had flat feet.[6] Therefore, it is not only modern humans that sometimes have flat feet. *Australopithecus afarensis* apparently also occasionally suffered from this condition. All the other features of the Trachilos prints point to a foot made for walking, which means the prints are clearly humanlike. Thus, the find from Crete contradicts the assertion that the human walking foot first developed 3.7 million years ago with *Australopithecus afarensis*. It is far more likely that upright gait in combination with a foot clearly made for walking developed much earlier, probably more than 6 million years ago.

How Exact Is the Molecular Clock?

The timing discussed above is supported by recent estimates made using molecular genetics. A few years ago, with the help of what is known as a molecular clock, researchers set the time when the evolutionary lines of humans and chimpanzees split at about 7 million years in the past. The molecular clock is based on the idea that mutations—that is, random changes in the genetic material of living entities—happen at a somewhat regular rate over long periods of time. Mutations are the motors of evolution. They change organisms, and these changes are usually passed down when they benefit the organisms that carry them. And that is how new species develop step by step.

To find out when the evolutionary lines of different species split off from each other, you have to look back and convert the differences in genetic material into time. The further back in time the split happened, the greater the difference. Initially, this system could only be checked and calibrated using fossils of a known age. For example, there are 13-million-year-old orangutan finds whose anatomical features show that they stand at the very beginning of orangutan evolution. The genetic material of humans and orangutans differs by at least 3 percent. Therefore, according to the pioneers of the molecular clock, the differences in genetic material must have accumulated over this period of time. According to this clock, the split between humans and chimpanzees must have happened later, because their genetic material differs by only 1.3 percent. Based on this reckoning, the gorilla evolutionary line split off from ours about 10 million years ago, between the time the evolutionary line of the orangutans split off and the time the evolutionary line of the chimpanzees split off.

The system was relatively rough, but for a long time it gave a good preliminary overview that agreed with what we knew from fossils we had already dated. Then, with the arrival of paleogenetics, this field of research took off. Suddenly, it became possible to extract and analyze genetic material from fossils of extinct species. This meant you could calculate mutation rates within species. Today, experts can determine the number of genetic changes between living members of a species and their ancestors. Paleogenetics cannot look back millions of years in the past, but it can look back many hundreds of thousands of years, time periods in which many mutations can accumulate in genetic material.

On the basis of these much more exact methods, scientists began to realize that mutations often do not happen at regular intervals and that the rate of mutation can fluctuate considerably. Even more important, mutation rates differ from species to species because they are influenced by a wide range of biological factors. For example, in modern humans, the age of the mother and the father, the quality of the sperm, metabolic rate, body mass, and population size all play a role. More recent approaches go even further and try to determine mutation rates by looking at genetic differences between parents and their children. This process obviously does not involve any fossils.

Advances in our understanding of how genetic mutation rates work suggest that human genetic material changes much more slowly than we thought. Using a combination of the molecular clock and paleogenetics, researchers have determined that in all probability, the split between humans and chimpanzees occurred earlier than the original estimate of 7 million years and likely up to 13 million years ago.[7]

There are, no doubt, many scientific papers yet to be written on this subject, but there is a definite trend emerging. Clearly, we need to look further back if we want to understand our early evolutionary history better. These results could also help explain why a biped was walking along a beach in Crete at least 6 million years ago. It seems almost ironic that Ahlberg's research team tried unsuccessfully for six and a half years to get their groundbreaking research published. Many scientific journals rejected the manuscript after anonymous reviewers read it, and their reasons were often difficult to decipher. After our article on *Graecopithecus* appeared in the summer of 2017, however, the dam finally broke.

– 12 –

A SKULL IN THE SAND AND A "SECRET" THIGHBONE

The Shady Case of *Sahelanthropus*

T HE SEARCH FOR the origin of humankind sometimes leads researchers to far-flung places. The Djurab Desert in northern Chad is a case in point. Here, fossil-bearing rocks lying close to the surface date from the time when many researchers believe the human and chimpanzee evolutionary lines split. And that is why the French paleontologist Michel Brunet has spent thirty years exploring the area.

It is an exceptionally inhospitable, extremely dry desert landscape in the central Sahara south of the Tibesti Mountains. Temperatures often exceed 122 degrees Fahrenheit (50 degrees Centigrade). Hot, dry winds sculpt the dunes, blowing almost constantly over the bare expanses of sand. It is the dustiest place on Earth. Up to 700 million tons of rock dust are stripped away every year.[8] The amount of dust

118

transported around the world from this desert annually is roughly equivalent to the Great Pyramid of Giza being eroded and blown away 109 times in a single year.

The French geologist Alain Beauvilain knows this isolated, lifeless area better than almost anyone. From January 1994 to July 2002, he coordinated the fieldwork of the Mission Paléontologique Franco-Tchadienne. His boss was Michel Brunet.

In July 2001, Beauvilain was once again out in the Djurab Desert with a team of three Chadian researchers. Brunet did not come along on this brief routine tour. The area the team was investigating, Toros Menalla, is known for its abundance of fossils. The bone hunters had been to the area many times before. Part of their protocol was to regularly revisit those areas where the researchers thought old sediments might appear. The scientists hoped that the powerful wind might do some of their tedious work for them. It came up every day, and it always blew from the same direction, constantly re-sculpting a landscape devoid of vegetation. Violent storms and squalls often conjured up and then blew away elongated sand dunes called ergs. During storms, myriad grains of sand would scour the ground like a sandblasting machine, revealing what lay beneath. If the sand contained larger, compact features—gravel, for instance, or bones—the wind would blow away the finer particles. Stones, scree, and sometimes even fossils were left behind as though the landscape had been passed through a sieve. These cleaned-out pockets could be found in many of the valleys between the ergs.

Members of many expeditions over the years had found traces of once-fertile ecosystems in these vast expanses of sand. From the numerous fossils they found, scientists confirmed the existence of savannah-like landscapes that extended along rivers and around lakes and were full of riparian forests. The remains of large fish were proof that there had also once been deepwater bodies supplied with an abundance of moisture. Land that is barren today had once teemed with life: amphibians, crocodiles, turtles, rodents, rhinoceroses, prehistoric pigs, giraffes, three-toed horses, and hyenas.

As valuable as these finds were, the French and Chadian researchers had been searching for years and had not been able to track down the real prize they were seeking: fossils of early hominins. Finally, in 1995, Beauvilain's team had some luck when it found a 3.5-million-year-old lower jawbone from a species of *Australopithecus*.[9] Since then, however, there had been no more exciting finds, which, of course, severely tested the staying power of the participants. Then, on the morning of July 19, 2001, their perseverance finally paid off.[10]

"Hope of Life"

A sandstorm threatened, so the four researchers parked their two pickup trucks up on the highest point of a dune. That way, they could find their way back to them more easily, even if the vehicles were enveloped in a cloud of dust. They paired off to search the depression below for fossils. After a while, one of the Chadian students employed by Beauvilain, Ahounta Djimdoumalbaye, noticed a number of unusual dark objects spread over about 1 square yard (1 square meter) of sand. A round shape about the size of a small bowling ball stood out. It didn't take him long to realize that it was a skull.

The skull was unusually intact. Only the lower jawbone was missing. The outside was encrusted with a black substance. The face was misshapen as though it had been squashed under a heavy weight, but it clearly had the features of an ape. Was this, finally, the sensational find they had been hoping for? The student felt his heart beating faster. Two years earlier, when he had first joined an expedition into the Djurab Desert, Brunet had spoken to him in French and told him, "I'm sure that if anyone is going to find a primate here, that person is going to be you." Was he going to have beginner's luck after all—just delayed by a couple of years? "We have what we seek!" he called to his colleague Fanoné Gongdibé, who was working not far away. "We are victorious!"[11]

Excited, they waved over another member of the team and called to Beauvilain that he should hurry and get the camera, because they had found a great ape or perhaps even an early hominin. At first Beauvilain thought it was a joke, but he realized the moment he saw the find that it was indeed something important. He took photos and recorded a video, and confirmed the GPS coordinates so they could locate the exact site again. By midday, the research team had found a further one hundred bones lying close by in the sands of the dunes. Most came from a variety of different mammals, and later the research team used them to estimate the age of the skull.

A year later, news of the sensational find spread like wildfire through media outlets around the world. Michel Brunet and his colleagues, with the blessing of the Chadian government, had first cleaned the skull and examined it thoroughly at the University of Poitiers, and then had it analyzed in detail using state-of-the-art tools at the University of Zurich. Their results were published on July 11, 2002.[12] The key message of those results: The oldest known ancient ancestor of humans came out of North Africa and lived from 7 million to 6 million years ago. It was, therefore, "close to the divergence of hominids and chimpanzees." The scientists named it *Sahelanthropus tchadensis*. Sahel is a geographic region at the southern edge of the Sahara, so the name translates as "the Sahel human from Chad." They gave it the nickname Toumaï, which in the local Dazaga language means "hope of life."

The skull from Chad had small canines, a relatively flat face, heavy protuberances over the eye sockets that met in the middle, and medium tooth enamel—all features that suggest a humanlike creature.

The collection of bones at the site where *Sahelanthropus tchadensis*'s skull was found on July 19, 2001. The arrow points to the thighbone belonging to this species. This bone has since disappeared.

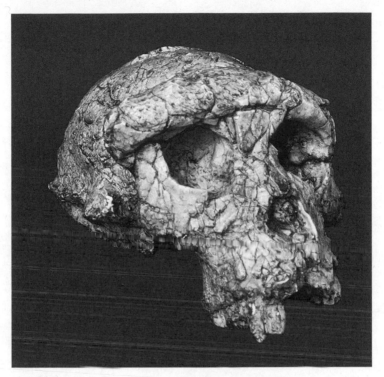

The skull of *Sahelanthropus tchadensis* after it was cleaned

It was not long, however, before the first critical voices began to make themselves heard. A few months after *Sahelanthropus tchadensis* was initially described, a Franco-American team led by Milford Wolpoff chimed in with an open letter to *Nature*.[13] The team included Brigitte Senut and Martin Pickford, who had discovered *Orrorin tugenensis*, which until the paper about *Sahelanthropus* was published was credited as being the earliest potential hominin. In this letter, the three scientists expressed doubts about how distinctive the skull's features really were. They argued that the features, although they looked as though they might belong to a humanlike creature, evolved independently and were not features that could be traced back to a common ancestry. The small canines suggested a female. The hefty bones of *Sahelanthropus* and especially the shape of the backside of the skull were reminiscent of a gorilla. From that, the authors concluded that Toumaï was a female gorilla ancestor. Brunet's research team promptly defended its position. The discussion in the journal went something like this: Wolpoff's criticism—*Sahelanthropus* is a great ape because its features are too primitive for it to be an early hominin. Brunet's reply— Of course Toumaï has primitive features because it is the oldest early hominin. The two were at loggerheads.

THREE YEARS LATER, Brunet's team reconstructed Toumaï's skull using computed tomography scans.[14] Although the skull was unusually intact,[15] it was partially fractured and misshapen, which made it difficult for even the best experts to evaluate it. However, using techniques such as CT scans, it is possible to create a virtual reconstruction that corrects deformations. The reconstruction based on the scans convinced Brunet's researchers that *Sahelanthropus* was not a gorilla ancestor and that its reconstructed foramen magnum (the hole through which the spinal cord enters the skull) lay beneath its skull, which is indicative of bipedal locomotion because it means that the head is balanced on the spinal column. A foramen magnum situated farther back is indicative of an animal that walks on all fours. The area around the foramen magnum in the original fossil, however, was severely damaged.

This analysis was also questioned by Wolpoff and his team.[16] The criticism explained in much more detail that the features in question were in no way sufficient to support the analysis, and that the back part of the skull and bone attachments for the neck muscles, in particular, did not fit the anatomy of a biped.

Our El Graeco team compared *Sahelanthropus tchadensis* with the somewhat older *Graecopithecus freybergi* (see the diagram on page 32). We determined that the tooth roots on Toumaï's lower jaw were much more primitive than El Graeco's. The root of Toumaï's canine was longer and the roots of its two premolars were completely split and not grown together like those of *Graecopithecus*. These features of the roots also suggest that *Sahelanthropus* was not an early hominin.

The Nape of the Neck in Great Apes, Sahelanthropus, and Modern Humans

■ Where the neck musculature attaches to the back of the head in chimpanzees

■ Where the neck musculature attaches to the back of the head in modern humans

C Direction of force in the neck musculature of chimpanzees

H Direction of force in the neck musculature of modern humans

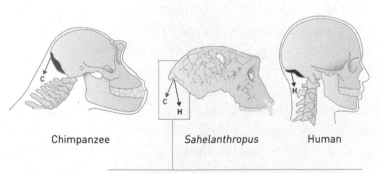

Chimpanzee   *Sahelanthropus*   Human

The human angle of attachment for the neck muscles (H) is unlikely for *Sahelanthropus*, because the angle this would create would be too tight for its neck muscles. The angle of the *Sahelanthropus* skull is more like that of a chimpanzee (C).

Despite these clearly well-founded arguments, many pale-ontologists continued to think it possible that *Sahelanthropus* walked on two legs and even that it was an early hominin. But in the end the puzzle was going to be solved only if more facts came to light. It would be of huge scientific significance if we were able to examine features of other parts of the body below the skull, as well, such as the spine or fragments of the leg bones. In the many articles written by Brunet's team, however, the only body parts mentioned are a skull, three fragments of a lower jawbone, and a few individual teeth. A few years after their publication, though, there was increasing speculation that other bones from *Sahelanthropus* existed.

The Case of the Disappearing Thighbone

Early in the summer of 2010, Martin Pickford, one of the scientists who was challenging Michel Brunet's conclusions, and I met with our colleague Roberto Macchiarelli at the University of Poitiers. Macchiarelli's laboratory was in the same section of the science building as Brunet's office. In passing, Macchiarelli showed us a photograph he had on his computer of a black bone. "That's Toumaï's thighbone," he told us. The research team had found it in 2001 right next to the skull (see the photograph on page 122), although neither then nor later was it ever suggested that it belonged to *Sahelanthro-pus*, perhaps because bite marks, probably from a hyena, had severely damaged both ends of the bone.

In 2004, a doctoral student called Aude Bergeret turned her attention to the complete collection of bones belonging to the sensational find as part of the work she was doing for her doctoral thesis. She noticed that she could not identify the main shaft of the bone as belonging to any animal—other than *Sahelanthropus*. Brunet was away on a research trip and could not be reached. And so, on the spur of the moment, Bergeret requested the assistance of Macchiarelli, who was at the time the head of the department of geosciences. Macchiarelli and Bergeret spent many days analyzing the precious thighbone and tentatively concluded that although it was indeed from a great ape, in all probability it did not come from an ape that walked on two legs. When that became known, the thighbone disappeared without a trace and the doctoral student lost her position at the university.[17]

When I met with Macchiarelli and Pickford in Poitiers, we made a detailed study of multiple photographs of the bone that allegedly belonged to Toumaï and compared its anatomy with that of a thighbone from a chimpanzee and a thighbone from *Orrorin*, a 6-million-year-old fossil discovered in Kenya in 2000 that might be an early hominin. Because of the year *Orrorin* was discovered, this species is sometimes also called Millennial Man. In contrast to *Orrorin*, which most scientists consider is very likely an early hominin,[18] the thighbone from Toros Menalla had an obvious bend along its longitudinal axis. That is not something you see in an early hominin that walked on two legs.

Photographs, of course, cannot replace a thorough examination of the real thing. But, to date, the enigmatic length of bone has been unavailable for further examination. According to Brunet, his studies on the bone are a work in progress. He wrote in a brief email quoted in an article in *Nature*, "Nothing to say before publishing."[19]

This does bring up the question of why an investigation into a bone remains unfinished almost two decades after its discovery. Whatever the reason, such an important find should not be kept from the scientific community.

Macchiarelli and Bergeret did not want to wait any longer. In early 2018, fourteen years after their first examination, they decided to open up discussion by giving a presentation on what they knew about Toumaï at the annual meeting of the prestigious Société d'Anthropologie de Paris. The committee, however, turned down their application—a highly unusual response, because there was great interest in what they had to say. Could it be that Michel Brunet, one of the icons of French science, Knight of the Légion d'honneur, recipient of the Ordre national du Mérite, did not want to be challenged? Was the upright gait of *Sahelanthropus* and therefore its status as an early hominin something that was not to be questioned? After all, there is a road in the university campus at Poitiers named after Brunet and the parking garage at the station bears the name "Parking de la Gare Toumaï."

The way information about *Sahelanthropus*'s thighbone is being handled is unfortunate for science. It is impeding the important process of independent review of the dates and of the statements that have been made about the find. These unscrupulous dealings by influential groups and local networks, as Macchiarelli has stated publicly,[20] are also bringing the well-regarded French paleoanthropological community into disrepute.[21]

There are more inconsistencies in the case of *Sahelanthropus*. The ancestry and the age of the find are anything but certain. As Alain Beauvilain wrote in an article,[22] Toumaï's remains were not dug out from the sediment where they originally lay. They were not found in a ground layer that could be assigned a firm date. The find, therefore, was not, to use a technical term, in situ. Yet that was the impression given in all the articles published by Brunet's team. Instead, the skull and the thighbone, along with many other bones from other species of animals, were lying on sand dunes likely formed by the strong desert winds only days before the discovery. Sand dunes are highly mobile. Some of them in the Djurab Desert move as much as 650 feet (200 meters) a year. This means that the objects within them are constantly being moved from place to place. A fact that, according to Beauvilain, also presents huge problems for military personnel clearing mines in the area.

This complicates things enormously for paleontologists and archeologists who do their research in deserts. They know dunes are not places where you can make conclusive finds, because on the one hand, you cannot date the finds accurately, and on the other, solid objects quickly move down to the base of the dunes because of their weight. That fact also informed the way the Mission Paléontologique Franco-Tchadienne conducted its searches. The researchers preferred to search the valleys between the dunes for fossils.

According to statements made by Beauvilain, Toumaï's skull and the thighbone the team found had no connection with any rocks that could be dated. Indeed, because some of the bones were oriented directly toward Mecca, the geologist even thought nomads might have assumed they were human remains and given them a Muslim burial.[23] They could, therefore, have come from anywhere within a wide area. Despite that, Brunet and his colleagues dated *Sahelanthropus tchadensis* with extreme precision in multiple studies, all of which confirmed their very first estimate of the age of the find back in 2002. The rock profiles they mention vary from article to article, although all are correlated directly with the place where the fossil material was "dug up."

It is a shame that these inconsistencies and the lack of open discussion overshadow the remarkable paleontological results from Toros Menalla. Was it all meant to defend the theory that the oldest early hominin came from Africa, no matter the price?

One thing that is not up for debate is that *Graecopithecus* is older than *Sahelanthropus*. The Saharan dust in which El Graeco's lower jawbone was embedded in Pyrgos originated in desert sediments that today lie buried underground at Toros Menalla.[24] Powerful winds transported the dust northward over the Mediterranean over 7 million years ago.

FROM EARLY HOMININ TO PREHISTORIC HUMAN

The Out-of-Africa Theory Begins to Wobble

I T IS DIFFICULT even for experienced experts to judge the reliability of the evidence we have about the earliest hominins. Quite a few facts suggest that the species *Orrorin tugenensis* has a high probability of being an early member of the human ancestral line. Whether *Sahelanthropus tchadensis* will be classified as a hominin in the future is more questionable, for the reasons given in the previous chapter. It is also difficult to assess *Ardipithecus ramidus*—found in Ethiopia and dated back to 4.4 million years—because this species has a grasping foot similar to that of a great ape rather than a humanlike foot designed for walking. The size and shape of the bones of *Ardipithecus* found so far suggest they belong to a female, and scientists have called her Ardi for short. Given her significantly splayed big toe as well as the particular shape of her grasping feet, her powerful legs, and the structure of her pelvis, the researchers who examined the find concluded that Ardi mostly balanced on branches while standing upright and occasionally walked on the ground on two feet. But there is not, as yet, sufficient evidence to prove this.

Even the scientific data we currently have from *Graecopithecus* and from the footprints at Trachilos are far from being completely convincing, but they do give us ample reason to question the current evolutionary model for our ancestral lineage.

The oldest species that is currently accepted without question as an ancient human ancestor is *Australopithecus afarensis*, which gained iconic status when Lucy's skeleton was found.[25] The reason this species is so enormously important for our understanding of early hominin evolution is that with more than four hundred individual finds, *Australopithecus* is by far the best-researched species of early hominin. The finds from the Afar Region in Ethiopia and Laetoli in Tanzania are especially numerous, and all the finds are between 3.7 million and 3 million years old.

From members of Lucy's kin, we have discovered a wealth of adaptations to bipedal gait in almost all parts of the body. The spine is curved in the lower back, the pelvic girdle is shortened and rotated forward. The thighbones angle inward and the knees are capable of full extension. The big toes are robust and closely aligned with the other toes. The foot shows signs of an arch to cushion the impact when the feet hit the ground. But there are also a few primitive features that distinguish Lucy and her kin from later prehistoric humans: relatively long forearms, short thighs, curved fingers and toes, and a specific kind of arm and shoulder musculature that can be reconstructed using the bones that have been found. Based on the specific shape of the hip joint and pelvis, many experts agree that Lucy and others of her species had not yet committed entirely to walking on the ground and still spent at least some of their time up in the trees.[26] Other researchers, however, see no anatomical constraints to walking and running.[27]

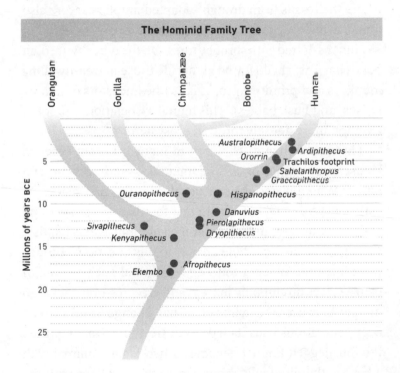

The Hominid Family Tree

The Advantages of Walking Upright

Australopithecus afarensis, when mature, had a brain volume of about 450 cubic centimeters. The relationship of its brain

volume to body mass was just a little more than a modern
chimpanzee's. The "southern ape from Afar" was an extremely resilient species that clearly adapted exceedingly
well to changing circumstances. The finds we have to date
prove this species existed for a relatively long time—more
than 700,000 years—even though the climate and the ecosystems in which they lived changed dramatically many
times. On the one hand, their fossils come from soil layers
where the fossils from drought-adapted antelopes were also
found, which suggests a habitat of wide-open grassland or
savannahs dotted with shrubby trees. On the other hand, their
bones have also been found alongside those of tree-dwelling
monkeys and primitive prosimians (the ancestors of lemurs,
lorises, and bush babies). This faunal association is proof of
a wetter, more densely forested ecosystem.

The mosaic-like distribution in many different types of
landscapes touches on a fundamental issue in human evolution: Why did upright gait develop in the first place? As
early as 1809, long before the first fossils of great apes and
early humans were found, the French naturalist Jean-Baptiste
Lamarck had an incredible answer to this question. His idea
went like this: When a four-legged animal such as a chimpanzee gives up climbing in trees and grasping branches with
its feet because its environment has changed and instead
over generations uses its legs exclusively for walking, then
the four-legged animal becomes a two-legged animal with
a big toe that gradually grows closer to and aligns with the
other toes.[28]

Over the next 210 years, generations of scientists, including Charles Darwin, transformed this theoretical musing
into the most influential and long-lived hypothesis about

human evolution: the savannah hypothesis. This hypothesis postulates that at a time when the climate was very dry, the ancestors of modern humans left the retreating forests to live from that time forward on the ground in a savannah-like landscape. It has since been expanded and refined, and also questioned and qualified, but its underlying ideas still apply.[29] Life beyond the trees in an open ecosystem dominated by grasses brought with it more challenges and dangers, while also offering new opportunities for survival and new sources of food. Some features proved to be more useful on the ground compared with conditions in the maze of the tree canopy. And so, the genetic changes selected for were those that allowed anatomical details, metabolism, and behavior to better adapt to an environment that was itself in a state of flux.

When you walk upright, you no longer need your hands for locomotion and so they can be repurposed to carry food, offspring, or useful objects. This means you can develop fine-motor skills. When your hands are free, you can fashion tools and conquer fire. Also, walking on two feet is more efficient than walking on four. It is true that a biped is not as fast as a leopard, but it can keep up a running pace for longer than an antelope and it can do so while carrying a weapon. Finally, walking upright was the best protection from overheating in a landscape with little shade, because it reduced the body area directly exposed to sunlight.

Despite these persuasive arguments, a few experts seriously question the savannah hypothesis. One reason for their doubts is the way some researchers interpret Ardi's form of locomotion. If this species did indeed walk upright in trees on two legs from branch to branch using its powerful grasping feet and relying on its arms and hands for additional support, the savannah hypothesis seems to lose its significance as an explanation for the development of an upright gait. At least, that is what some experts think. It is therefore unsurprising that Tim White, one of the researchers who discovered *Ardipithecus ramidus*, is the main opponent of the savannah hypothesis and sees the origins of bipedalism in Ardi's anatomy.

However, an ape that weighs about 110 pounds (50 kilograms) and is about 4 feet (1.2 meters) tall and balances upright on branches is not that unusual. The much-older *Danuvius guggenmosi*, our Udo, the ape from the Allgäu, moved in a very similar manner long before Ardi existed. But *Danuvius* likely could not last for long distances walking on two feet on the ground. There is therefore no compelling logical argument that the upright gait practiced by *Ardipithecus* with its grasping foot was necessarily a first step in the direction of highly developed bipedalism with a foot designed for walking. The footprints from Trachilos, which are more than 6 million years old, are also evidence against this hypothesis as they predate Ardi by almost 2 million years. This means the walking foot that sets humans apart developed a long time before *Ardipithecus*. Ardi's particular form of locomotion is therefore not sufficient to refute the savannah hypothesis. In my opinion, *Ardipithecus* was an evolutionary side branch that did not stand at the beginning of bipedalism.

The savannah concept is also compelling because it places changes in the environment and the climate as central factors driving evolution. Yet, the farther back we push the dates for early hominin fossils in Africa and the better we get at making detailed reconstructions of the climate history of Africa, the clearer it becomes that by the time savannahs spread over large areas of that continent, the first steps to walking on two feet had already been taken by early hominins. To hold onto the idea of Africa as the cradle of humanity, a few influential paleontologists finally threw away the savannah hypothesis and argued that in the beginning, perhaps only small populations of great apes living on the margins of the African tropics developed upright gait or maybe upright gait developed from walking upright on branches. The possibility that the savannah hypothesis was correct, but locating the home of our earliest ancestors in East Africa was not, seemed unthinkable.

While we were determining that the fauna from Pikermi was about 7.3 million years old and that *Graecopithecus* from Pyrgos was about 7.2 million years old, it became clear that savannahs spread far earlier in Europe and the Middle East than they did in Africa, where they first appeared 2.6 million years ago. If you accept that the split between the human lineage and the chimpanzee lineage happened not in Africa but in Eurasia—as both *Graecopithecus* and the Trachilos footprints suggest—then the date for the final divergence—around 7 million years ago, as calculated using genetic analysis—suddenly lines up once again with the arrival of savannah landscapes. (More on this in Chapter 15.)

Great Ape Evolution, Global Climate, and Ecosystems

First phase ca. 21–14 million years ago

Primitive apes arise in Africa.

Globally very humid and warm, 9°F (5°C) warmer than today. Tropical temperatures in Europe and little ice in eastern Antarctica.

Forests into the polar regions. Many evergreen trees in Europe. First deserts begin to appear in Central Asia.

Second phase ca. 14–7 million years ago

Prime time for the hominids. They arrived in Eurasia and evolved into the groups we know today: orangutan-like, gorilla-like, chimpanzee-like, and the earliest hominins.

Globally very warm, 5.4°F (3°C) warmer than today. Subtropical temperatures in central latitudes. Only eastern Antarctica is completely covered in ice.

Forests yield to open savannahs in many areas. Deserts and steppes expand in Asia.

Third phase

ca. 7 million years ago until today

Great apes died out in Europe and migrated to Africa, where early hominins flourished. At the beginning of the ice age, the first early humans appeared at almost exactly the same time in both Africa and Eurasia.	Highly variable climate conditions. Fluctuations between warm periods and glacial periods. Antarctica and, increasingly, the Arctic completely freeze over.	Globally, grasslands (savannahs, steppes) expand. Deserts dominate wide expanses of the Old World. Airborne dust increases. 5.5 million years ago, the Mediterranean dried up.

Out of Africa?

The finds of what appear to be early hominins in Europe, finds that fit with the timeline of savannahs appearing in Europe, clearly contradict the assumption that Africa is the cradle of humanity. In the 1980s, the idea that the human evolutionary line arose in Africa and that humans spread out all over the world from there was labeled the Out of Africa theory. The name comes courtesy of Günther Bräuer, a long-time professor of anthropology at the University of Hamburg, who was inspired by a film based on Karen Blixen's memoir, *Out of Africa*.

Today, proponents of this theory propose that there were two chronologically and geographically distinct migrations: Out of Africa I describes the spread of early humans more than 5 million years after El Graeco. Out of Africa II refers to the later spread of modern humans, *Homo sapiens*.

Whichever idea about the development of humans you prefer, what is certain is that the evolution of the genus *Homo* is closely tied to the dramatic changes in climate ushered in by the ice age that began on our planet about 2.7 million years ago. Around the onset of the ice age, there were profound advances in human evolution, and the oldest fossils of the genus *Homo* come from this time. However, they left behind more than just their bones—they also left behind the tools they made.

To date, very few fossil bones from the first early humans between 2.6 million and 1.9 million years old have been found in Africa. Whether the few finds we know of from Ethiopia, Kenya, and Malawi[30] belong to a single species or to many different species has not yet been fully resolved.[31] Their familial relationship with *Australopithecus* is also the subject of vigorous scientific debate.[32] To avoid academic confusion about terminology, paleontologists have simplified things and call them "early *Homo*" without referring to species names. The first early *Homo*, or early humans, are differentiated from early hominins by their round rather than oval dental ridges and, above all, by their larger brains, with a volume about half the size of the brain of a modern human. A substantially larger brain indicates a different nutritional base and is certainly an important requirement for more complex cognitive processes, which are necessary for making tools, for example.

Early hominins also likely used tools, as modern chimpanzees do today. But early members of the genus *Homo* not only used whatever objects they found lying around as tools, they also purposefully made tools for specific tasks, and these tools, in scientific terms, are called artifacts. Pebble tools are the defining characteristic of the oldest early humans, and groups that used early techniques to make these stone tools are called Oldowan cultures. The earliest pebble tools, found by Mary and Louis Leakey in the Olduvai Gorge, were choppers. A chopper is a stone that has been hit with another stone to remove flakes and create a sharp edge. The oldest examples in East Africa are 2.6 million years old and therefore fit chronologically with the first skeletal remains of the oldest early humans.

Thanks to ever more precise dating methods, it is possible today to get an increasingly clear picture of events at the beginning of the ice age. There are, however, still many unanswered questions. It is not easy, for example, to recognize primitive tools as tools. Rocks manufactured by humans are difficult to distinguish from lookalikes generated by natural processes. On the beds of mountain streams, for instance, stones can knock against each other, splintering off fragments that look remarkably like Oldowan tools. In such cases, the only way to prove sharp rocks are tools made by humans is to perform a detailed analysis of the geological origin of the rocks and their location relative to fossil finds.

For decades, no one questioned the Out of Africa I theory. According to this idea, early humans belonging to the tool-manufacturing genus *Homo* arose over 2 million years ago in East Africa. Then, well over a million years later, after *Homo erectus* (Upright Man) developed, early humans conquered Eurasia. However, new discoveries of tools found outside East Africa, in the Mediterranean region[33] and Asia, shook the Out of Africa I scenario to its core.

Incontrovertible Traces in Asia

Sensational finds from India played a big role in undermining the idea that tool-making cultures arose solely in Africa. The oldest handmade tools from Asia were found by scientists in Masol in the Punjab in 2016.[34] Analysis revealed that the tools had been used 2.6 million years ago. If this is correct, they are the oldest tools used by early humans in Asia and at least as old as the Oldowan finds in Africa. This news upset the international scientific community. Many scientists either did not react at all or reacted very cautiously. But the evidence the French and Indian researchers found was compelling. On steep slopes in this mountainous region of India, they discovered about 250 tools, as well as a large number of fossil bones that looked as though someone had used tools to work on them. Because the first fossils the scientists found were not embedded in sediment they could date with confidence, they dug deeper—with success. They found more bones and tools in material that they were able to date.

New research results from China shook the Out of Africa I theory even more. In recent years, strong evidence has accumulated that there have been human settlements in China for far longer than 2 million years. The best-researched location in China is the Longgupo Cave in the 6,500-foot-high (2,000-meter) karst mountains in Sichuan province. The cave lies in a thick mountain forest 2,460 feet (750 meters) up the steep slopes of the raging Yangtze River. Chinese and French scientists dug down 40 feet (12 meters) in the massive 65-foot (20-meter) layer of sediment on the cave floor and found thousands of remains of a rich fossil fauna, including giant pandas, elephant-like stegodons, and saber-toothed cats. The researchers discovered tools in a total of twenty-seven levels. As they dug down into the sediment in the cave, they came across artifacts practically every few feet. In the end, they carefully removed over a thousand tools—a variety of simple stone tools and rocks used for pounding in the Oldowan style.[35] When the oldest and deepest layer containing artifacts was dated, it was discovered to be 2.48 million years old. And the excavations have not yet reached the lowest 26 feet (8 meters) of the cave sediment.

The researchers also unearthed remarkable remains of great apes from the cave sediments, including a handful of teeth from a gigantic ape that stood over 6 feet 6 inches (2 meters) tall, a distant relative of modern-day orangutans.[36] There were humanlike teeth there, too: a fragment of a lower jawbone with two molars and an upper incisor. Chinese scientists ascribed both fossils to the newly described early human species *Homo wushanensis*, also known as Wushan Man.

North American researchers got the chance to examine the finds in 1992.[37] They came to the conclusion that Wushan Man was an early *Homo*, a species in the human genus, generally confirming the results of their Chinese colleagues.

That sent shock waves through the classic Out of Africa I theory yet again. First, according to this theory, no ancient early *Homo* had ever left Africa. The only species to leave had been a highly developed early human. Second, after the discoveries in China, Africa no longer had a head start in terms of the time the genus *Homo* arose. The finds from Asia were from 2.6 million to 2.48 million years old, which made them almost exactly the same age as the finds from Africa. In 2009, however, one of the American scientists abruptly retracted the conclusions he had put his name to fourteen years earlier. In an essay in *Nature*, he admitted that he had made a mistake,[38] something that rarely happens in science. Instead, he characterized *Homo wushanensis* as a "mystery ape of Pleistocene Asia," completely ignoring the tools that had been found. He cited the primitive shape of the teeth as the main reason for his retraction.

It is true that the teeth of Wushan Man are relatively small. They are about the same size as those of Udo from the Allgäu, who was only 3 feet (1 meter) tall, and the roots look slightly more primitive than those of El Graeco from the Balkans. However, *Homo luzonensis*, a species of early human recently discovered in the Philippines that is described in more detail in Chapter 19, has similarly small teeth with primitive-looking roots. In addition, these features of the teeth were noted in the 1995 publication and they lie within an accepted range for early hominins and for the oldest early humans. The excavations in China also provided a lot more "watertight" evidence for Wushan Man. So, why the spectacular retreat? Was it to avoid jeopardizing the Out of Africa I hypothesis?

It's possible that paleoanthropological discoveries between 1995 and 2009 did not fit with the accepted picture and that some scientists found the new finds unsettling. Modern dating methods, for example, have corrected the age of the oldest early humans in Asia many times, pushing them further and further back into the past.

In 2004, on the Indonesian island of Flores, a human of small stature with extremely primitive features stepped into the public spotlight: *Homo floresiensis*, aka the Hobbit. This species delivered another blow to the picture of human evolution centered on Africa, because the Hobbit is anatomically more primitive than the archetypal Asian early human *Homo erectus*, which, according to the currently accepted theory, was the first species of *Homo* to leave Africa.

The Paradigm Crumbles

Today, there is no longer an appreciable time difference between the oldest early humans and tool-making culture in East Africa, the Mediterranean region, and Asia. In both Africa and Eurasia, the evidence stretches back to the beginning of the ice age 2.6 million years ago. This makes the Out of Africa I theory untenable.

That means the popular idea that the Eurasian early human *Homo erectus* developed from the East African species *Homo ergaster* is also questionable. Today, it seems much more likely that both species, which are anatomically very similar, lived at the same time. The well-preserved skeleton of the African early human *Homo ergaster* (Turkana Boy) is dated to 1.55 million years ago, and a few other, albeit badly preserved, bone fragments suggest that this species of human might have settled the area around Lake Turkana as long as 1.7 million years ago. The archetypal Asian early human *Homo erectus* from the Indonesian island of Java has been dated to 1.6 to 1.5 million years ago.[39] Colleagues have even dated teeth from *Homo erectus* in China to as far back as 1.7 million years ago[40] and a skull from *Homo erectus*, also from China, to 1.63 million years ago.[41]

Apart from that, the anatomy of these early humans does not definitively suggest any shared ancestry between the two species. Neither do the tools they both used. Unlike the very early species of *Homo* with their simple Oldowan stone tools, both these species were using considerably more refined tools. Their tool technology, with its variety of carefully crafted handaxes, is known as the Acheulean culture.

The oldest of these handaxes, which scientists found at Lake Turkana in Africa, are 1.76 million years old.[42] At the moment, we do not have any handaxes from locations in Asia where *Homo erectus* has been found. However, on the Taman Peninsula, which lies between the Sea of Asov and the Black Sea, researchers have found a tool tradition that bridges the gap between the Oldowan and Acheulean cultures and can be dated back 1.6 million years.[43] The traditional Out of Africa I theory cannot explain the simultaneous appearance of early *Homo* and the archetypal early humans in Eurasia and Africa. It seems the attempt to restrict the origin of humankind to only one country, region, or continent has failed. Africa clearly was not the only cradle of humanity. Asia and Europe also had a big role to play in how we became human. Therefore, paleoanthropology urgently needs new hypotheses so that the data we now have can be categorized in a way that makes sense.

Perhaps new attempts at an explanation will build on the multiregional origin model put forward by Franz Weidenreich, who suggested back in 1943 that "Human evolution was not limited to a specific geographic center, but went on over a vast area comprising, possibly, the entire Old World."[44]

As explained, it is difficult to pin down where in the world a fossil species arose, partly because many species settled in regions of undetermined size, and only in rare cases can we directly study the processes by which species arose. Most fossil mammal species appear in the locations as though they were just passing through. That is mostly because fossil sites and excavations can give us only fleeting snapshots of the time and place where these species lived. It is only when you have a large number of these fleeting moments that you can assemble the snapshots into an album that makes sense of the whole.

Oldowan Stone Tools and Acheulean Handaxes

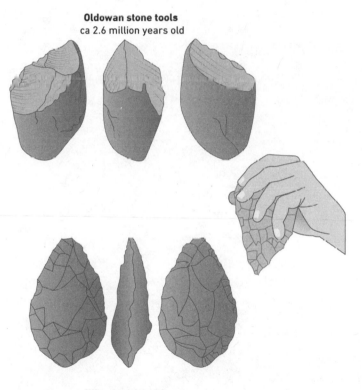

Oldowan stone tools
ca 2.6 million years old

Acheulean handaxes
ca. 1.76 million years old

Part 4

CLIMATE CHANGE AS A DRIVER OF EVOLUTION

– 14 –

NOT JUST
COUNTING BONES

Reconstructing the Environment Is Key

———

RESEARCHING HUMAN EVOLUTION is a story of big digs and fascinating discoveries. Interest is almost always centered on finding particularly spectacular fossils—a skull that looks ancient, perhaps, that gives us the feeling we are literally face to face with our ancestors. Researchers often focus on finding early hominin and early human fossils and examining them themselves. That is understandable, because it is only through these finds that we can figure out how the anatomy of our ancestors changed over time, and these fossils are the only places where we can find ancient genetic material, one of the most important sources of information in modern paleoanthropology.[1] To really understand the evolutionary process, however, you need to track down far more than fossils. Teeth and bones are only one part of the ancient puzzle. How do paleontologists and paleoanthropologists reach their conclusions? What methods do they use to reconstruct the world in which our ancestors lived when all they have to guide them is a few fossil finds that are missing most of their pieces?

At the center of every investigation lies a thorough investigation of the age of the finds. Without an exact date, you have nothing. Here is a striking example to illustrate the numerous snares that can trip you up when you are trying to determine the age of a find. Let's assume that over a period of two thousand years in the Pliocene—the epoch in which Lucy lived—there were six momentous natural events: an asteroid hit the Earth, a large volcano erupted, a giant desert was formed, our ancestors suffered a severe population decline, the remaining early hominins had to find a new place to live, and many large mammals went extinct. Even if we could date these events to plus or minus one thousand years—and we're a long way from being able to do that—it would seem to us that these events all happened at the same time. It would now be up to "scientific creativity" to reconstruct a chain of cause and effect between the events.

An attempt to reconstruct a prehistoric global catastrophe might go something like this: After an asteroid strike that triggered a large volcanic eruption, there were chaotic changes in climate around the world. As a result, living conditions in many regions deteriorated and numerous species of animals died out. Even our early ancestors had to fight to survive. A desert spread over the region where they used to live and many of them succumbed to this environmental crisis. But the survivors migrated to other regions, where they settled and thrived. This scenario sounds plausible. However, because the techniques we use for dating are not yet exact enough for us to separate out a few hundred years in the

Pliocene, it is equally possible that these events had nothing to do with one another.

To prove this, let's now look at a similar period of time in the past two thousand years in which six similar scenarios did indeed take place independently of one another. In June 1908, the Tunguska asteroid exploded above Siberia. In 1816, after the eruption of the Indonesian volcano Tambora, Europeans experienced a "year without a summer." There were catastrophic crop failures, and hunger pushed many people to emigrate to North America. In the fourteenth century, about one-third of the population of Europe died during the Black Death. In the ninth century, lions died out in Europe; in 1627, aurochs went extinct; and in 1936, the last of the Tasmanian tigers died.[2] From the fourth to the sixth centuries, entire tribes left Central Asia and migrated to Europe. About two thousand years ago, after a long period of drought in North Africa, the Sahara Desert as we know it today was created. All these events can be dated with precision and most of them are recorded in contemporary written documents; however, there is no causal connection between any of them.

Does this mean we can never make reliable statements about phases of evolution that happened millions of years in the past, because the lens through which we view the past inevitably distorts the picture and the rate of distortion increases the further back in time we travel? Actually, the situation is not that dire, because today we have at our disposal a wide range of methods we can use to reconstruct dates and environmental conditions even way back in the past, and not all significant events happen as closely spaced in time as the ones in the example above. Climate changes that are not triggered by natural catastrophes often occur slowly. What methods, then, do researchers use to decode the When, How, and Why of human evolution?

Sophisticated Dating Methods

The most important methods of dating include what are known as radiometric techniques. Radiocarbon dating is the radiometric technique most people have heard of.[3] It is based on measuring carbon in fossils or rocks. Carbon is a chemical element with the symbol C. Its nucleus usually contains six protons and six neutrons, which is why people refer to ^{12}C. Earth's atmosphere, however, also contains a carbon variant (a radioactive isotope of carbon) with eight neutrons instead of six, ^{14}C. It is constantly being formed when cosmic rays hit the atmosphere. The weakly radioactive ^{14}C, however, is unstable and degrades into its component parts at a steady rate. Living plants regularly absorb ^{14}C out of the carbon dioxide in the air when they photosynthesize, and they incorporate it, along with ^{12}C, into their tissues. The relationship between ^{14}C and ^{12}C in plants and in the animals that eat them therefore remains constant. After an organism dies, however, a kind of atomic clock begins to run. Because only the amount of ^{14}C present in the tissues at the time of death remains and no new ^{14}C is added, the portion of ^{14}C in the tissue declines at a constant rate through radioactive decay. This process is completely uniform, and it is therefore easy to translate it into time and use it to calculate the age of things.

The main restriction with radiocarbon dating is that because ^{14}C degrades relatively quickly, the portion of the isotope in fossils is reduced by half every 5,730 years. After about fifty thousand years, there is not enough of it left to date things precisely.

For this reason, paleontologists who want to look even further back into the past work with other radiometric methods as well. Fossils and rocks can be dated back 500,000 years with a high degree of accuracy using the decay rates of uranium and thorium. People also talk about the longer half-life of uranium. If you use other methods—for example, uranium–lead dating or potassium–argon dating—you can go even further back in time. But radiometric measurements are not always an option.

If the finds or the rocks in which they are embedded do not contain the necessary elements, then scientists must reach for other dating methods, such as magnetostratigraphy, which measures the orientation of tiny magnetic particles in rock. These particles are a permanent record of the direction of the Earth's magnetic field at the time they were laid down, acting like a compass etched in stone. Because the Earth's magnetic field is always moving and often completely reverses itself over the course of millions of years, this method delivers a precise window for the age of the rock and the fossils it contains. We used magnetostratigraphy, along with other techniques, when we calculated the dates for *Graecopithecus*.

Other methods rely on cyclic processes that run in precise known rhythms. These could be the rhythms of day turning into night or of summer being followed by winter, but they could also be fluctuations in solar radiation. Gravitational forces in our solar system cause cyclic variations in solar radiation at every point in the Earth. In lower and central latitudes, this cycle lasts twenty thousand years. In higher latitudes, nearer the poles, the cycle is twice as long. Because solar radiation is an important driver of climate, natural variations in climate are often cyclical. And these natural climate fluctuations are reflected in the characteristics of rocks. For instance, different sediments form in warmer climates than in cooler conditions. Therefore, the climate-dependent characteristics of rocks can be used to determine their age.

We have a wide range of dating methods at our disposal these days. If they are used correctly, they reduce the margin of error in our age estimates to less than 1 percent, which means an event that happened 1 million years ago can be dated to within 10,000 years of its occurrence. Or, to put it another way: events that happened yesterday can be pinned down to within a quarter of an hour.

Tiny Traces of Huge Happenings

Being able to date a find with precision is not enough to be able to place a fossil where it belongs in the evolutionary record. It is equally important to reconstruct the environment of the time using other organic and inorganic traces in the rocks. Of course, the fossils themselves are especially large and conspicuous organic remains. However, numerous other elements in the rocks contain useful information that is often overlooked by researchers. Tiny microfossils, for example, often tell you more about the past than large bones: the remains of unicellular algae, for instance, give us information about water quality in past times, and temperature, salt concentrations, and the direction of the current in ancient oceans are documented in the tiny calcareous skeletons of animal organisms such as plankton. Even organic molecules can be detected in rocks these days. Often, they are the only traces left of the earliest life-forms on Earth.

Fossilized plant pollen is another excellent source of information. It can help us not only identify individual plants but also reconstruct entire vegetative communities in lost ecosystems.

If the delicate organic pollen of plants is not saved in sediments, we can also use phytoliths, the microscopically small quartz particles that most plants store in their cells. Phytoliths are more resilient than pollen and often survive better over long periods of time. The burned remains of plants are also often very revealing. For instance, even tiny pieces of charcoal allow us to figure out how often a region was visited by fire over the millennia and which plants fell victim to the ancient infernos.

Organic remnants on early tools or containers used by humans can also tell us a great deal. We can read from them whether our ancestors used a stone tool to grind grain or butcher meat, and whether a clay pot held blood, milk, honey, wine, or even cheese. Usually all we need is a small shard from the pot.

If you examine the inorganic components of rock more closely, another window to the past opens. The size, shape, and layers of individual particles show, for example, whether the material was laid down in a sea or in a lake, whether the wind blew it in from a desert or whether it originally formed on the floor of a thick forest, a raging mountain stream flushed it down, or it was carried to its final resting place suspended in a sluggish current. Even the specific scratch marks that appear on the surface of minute particles of dust when you put them under a microscope tell you whether the dust was once caught up in the powerful winds of a storm—as we shall see in Chapter 15.

The chemical composition of the mineral components of the rocks also gives us valuable information about the paleo-environment. For example, the metabolic processes of bacteria in damp ground can attract tiny particles of magnetic iron oxide that, when measured with a magnetometer, give us information about the intensity of rainfall in those times. Then, if we examine the relationship between different oxygen isotopes in the mineral calcite, we can determine the ground temperature and the chemical composition of rainwater in geological epochs far back in the past. Even information about the altitude of the area in which the minerals were deposited is stored there.

All these methods have dramatically changed the way we carry out scientific digs today. The framework of the find—that is, how the fossils are embedded in a geological context—is crucial and must be precisely documented. Before a fossil is dug from the ground, its position is measured with a tachymeter.[4] Apart from that, we take numerous core samples from the rocks and do a detailed examination of the relationship of the layers both above and below the find. We also take photographs to document the condition of the fossils in situ, or on-site.

Fragile finds are often encased in a cast while they are still in the ground and later carefully freed from their protective covering once they are safely in the laboratory. New techniques, such as laser scans, allow us to recreate the find and its location digitally—as the researchers did with the footprints from Crete. Every detail, every observation from a dig could turn out to be important. Anything that is not documented is lost forever, because the excavation necessarily destroys the framework of the find. On a serious dig, the adventurous thrills associated with fossil-hunting expeditions are usually few and far between.

Every Detail Counts

When it comes to how modern technology helps with digs, paleontologists, scientists who study extinct animals, could learn a lot from archeologists, scientists who work with cultural relics. Archeological excavations are often carried out much more carefully than their paleontological counterparts. Many fossil collections in museums and research institutes leave you with the impression that our ancestors and their relations left behind nothing but teeth, along with the occasional jawbone. One of the reasons for this is that teeth covered with hard enamel are more resistant to decay than bones, which are softer, and therefore teeth survive better over long periods of time. I estimate that we have skeletal bones for only about one-quarter of the hundred or so known species of fossil humans and great apes.

In my experience, however, the reason for this gap in the fossil record is often inadequate fieldwork. I like to compare the current state of paleontological fieldwork with that of archeology in the nineteenth century, when it was all about bringing home treasure in the style of the German archeologist Heinrich Schliemann, who is famous for discovering the ancient city of Troy.[5] There's a reason the term "fossil hunter" is often used to describe paleontologists. And just as hunters choose their prey subjectively, for example, seeking out only

six-pointers in a herd of deer, fossil hunters, too, end up finding only what they are looking for—or nothing at all.

When things are done this way, the only finds that count
are trophies, and different bones are valued differently. A
tooth brings more prestige to the finder than a bone, an ape
fossil is more important than a fossil from a horse, a mammal
is more important than a fish or a plant. Rocks and sedimentary structures are relegated to the bottom of the hierarchy.
This subjective selection makes later analysis impossible.
And it means crucial bones from the lower body are often
overlooked.

With Udo, our ape from the Allgäu, we not only recovered
a shinbone, an elbow bone, and vertebrae, but also a kneecap, tiny bones from the wrist, and finger bones. Usually
you cannot classify small bones or bone fragments like this
on-site. Only after they are cleaned in the laboratory and
examined in detail is it possible to say whether a small bone
comes from the ankle of a deer or the wrist of an ape, and
whether the kneecap and toe or finger bone might come from
a predatory animal. We did not know that the large elbow
bone and the shinbone belonged to Udo until we got the
bones back to the laboratory and had a chance to examine
them in detail.

In Hammerschmiede we even found the hip bones from a
newborn baby elephant, the first find of its kind in the world.
The fossil is no bigger than the palm of your hand, and when
we dug it out, we thought it was part of a shell from a turtle.
Even so, we carefully salvaged the fragile bone. Now it, along
with other parts of the baby elephant skeleton, is an important source of information about the biology of this species
and possible causes of death.

And so, I have a few central principles when I conduct a fossil dig. The most important one is certainly this: There is no such thing as a good fossil or a bad fossil. Every fossil should be excavated and documented with care. And every observation of the dig site is potentially important and should be documented. Thorough digging should take priority over an extensive search or "hunt." The detailed reconstruction we have been able to make about the environment in which *Graecopithecus* lived shows how multifaceted and revealing our picture of evolution can be if we adhere to these principles.

BURIED IN THE
SANDS OF TIME

Landscape and Vegetation
in El Graeco's Time

L ET'S EMBARK ON a mental journey back into the distant past and take a walk. It's a spring afternoon about
7.2 million years ago and we're standing on the rocky
outcropping where the Acropolis will be built. It's 86 degrees
Fahrenheit (30 degrees Centigrade), and a light breeze
is blowing in from the coast. Standing under a bright blue,
cloudless sky, we have a spectacular view of the basin where
Athens will one day lie. At our feet a grassy plain stretches
out. Here and there individual trees and bushes rise up out
of the grass. Dark streams edged with thick cattails mean-
der through the landscape in the direction of the coast. The
land rises in the distance. A range of mountains formed of
limestone and marble rises to 3,200 feet (1,000 meters)
and surrounds the basin on three sides. A few pines dot the
sparsely vegetated slopes, their barrenness a stark contrast

to the landscape below. The foothills descend to a fringe of
thick oak forest.

The varying terrain and forms of the vegetation are so
harmoniously arranged over a relatively small area that it's
almost as though we're looking out over a landscaped park.
Suddenly the breeze freshens and the light takes on a reddish hue. To the south, in the direction of the sea, threatening
clouds are gathering, heralding an incoming storm. Time for
us to descend from the mountaintop and take a quick look
at the land below. We arrive at the foot of the rocky outcropping, where we find numerous springs. Impermeable layers
of slate lie under the limestone mountain, interrupting the
flow of rainwater that seeps down into its cracks and crevices.
The water flows out where the layers meet, creating a natural
source of water even in times of drought.[6]

So much water flows out of the springs that the area
around the base of the mountain has turned into a swamp.
To cross this soggy ground, we follow a largish stream that
starts in a cave. We disturb a giant hyrax, but the harmless
herbivore lumbers off, snuffling around in the lush growth
of the swamp. It's an ancestor of the African tree and rock
hyraxes that are alive today. They look like marmots, but they
are more closely related to elephants and manatees. While
modern hyraxes are no larger than rabbits, this ancient Greek
giant hyrax is the size of a pig.[7]

We wade farther through the shallow water of the stream.
Cattails are growing so densely on its banks that there's no
way we can walk through them. The farther we walk into the
plain, the thicker and more impenetrable the cattails become.
What might the landscape look like on the other side of the
reed beds?

As we stop to catch our breath, we become aware of unusual noises. There's a rustling and grunting close by. That can mean only one thing: pigs. Luckily, they haven't noticed us, because a mature *Microstonyx* boar is larger than today's wild pigs and could easily become extremely dangerous. Clearly, the animals are far too busy plowing their way through the soft ground for tasty reed roots to notice us. We proceed cautiously, and finally, after about another mile, we arrive at a ford across the river.

Elephants Larger Than Mammoths

This is a place where animals regularly come to cross the stream or to drink. The muddy banks are trampled clear of vegetation and completely covered in footprints. Our attention is immediately drawn to large, round prints about 20 inches (0.5 meters) across. There's no doubt what these are. They belong to the largest mammal alive at this time, the huge *Deinotherium proavum*. Standing more than 13 feet (4 meters) at the shoulder and weighing more than 15 tons, deinotheres were considerably larger than modern African elephants or ice-age mammoths.[8] They had two downward-curving tusks up to 3 feet (1 meter) long growing from their lower jaw, one on each side, which they may have used to root bulbs out of the ground. Most of the tracks we find, however, come from ungulates. Gazelles have left delicate prints barely more than an inch (3 centimeters) long. The imprints from the many antelope that live here are more than twice that size. Over ten species of antelope inhabit the Athens Basin, including a few with impressive spiraling horns.

Not far away, at the edge of the cattails, a small herd of roan antelope is grazing, unconcerned by our presence. Their compact bodies and long, saber-like horns are reminiscent of today's oryx. They're not at all disturbed by the many *Hipparion* horses alongside them, also filling their stomachs with fresh grass. From a distance, you could mistake hipparions for donkeys or gray zebras. But as soon as you see their prints in the mud, you can tell what they really are. Unlike today's horses, donkeys, and zebras, which walk on a single hoof, hipparions have two side toes in addition to the sturdy toe in the center. These side toes help stabilize them but mean they cannot run as fast as horses or zebras.

Has *Graecopithecus* perhaps also left tracks at this ford? We would love to know the shape of its feet and whether it was adept at walking on two legs or moved more like a modern chimpanzee. We search for a while without any luck. But perhaps we can find other evidence of El Graeco's presence—a broken branch, maybe, a collection of rocks, or a bone that has been smashed open? We concentrate on looking at the ground, sweeping our gaze from side to side, leaving no inch unexamined as we advance. Step by step, we distance ourselves from the stream. We're concentrating so hard, we don't even notice how far we've walked.

Suddenly, we are rudely awakened from our search. There, no more than 100 feet (30 meters) away from us, stands a loudly snuffling, natural lawnmower, barely taller than 3 feet (1 meter) with an elongated, cylindrical body and extremely short, equally cylindrical legs. Its overly wide mouth makes the primeval animal look as though it's grinning right at us. Two knife-sharp tusks, one in each corner of its mouth, underscore its grotesque appearance. Clearly, we've disturbed a grazing *Chilotherium*, otherwise known as a short-legged rhinoceros. Thanks to its short legs, its head is just above the ground, the ideal height from which to mow down grass with its blade-like tusks and rough coarse tongue.

Two other species of rhinoceros are living in the Athens Basin along with *Chilotherium,* and both are relatives of the modern black rhinoceros. Unlike the hornless short-legged rhinoceros, they each carry a single horn on their nose. Unfortunately, they do not show themselves today. Instead, we see, off in the distance, a pair of short-necked giraffes with their young. These are *Paleotragus*, precursors of modern okapis. The animals stand about 6 feet (2 meters) tall and mostly eat herbaceous plants and leaves from low-hanging branches. There are other giraffes in the Athens Basin that specialize in browsing the higher branches. The long-necked giraffe, *Bohlinia attica*, which grows up to 13 feet (4 meters) tall and looks similar to a modern giraffe, is good at harvesting tender new growth and acorns from the tree canopy using its 5.5-inch-long (14-centimeter) tongue. It's amazing how many herbivores live here. So far, we've seen no sign of the carnivores that hunt them.

The only meat-eaters we spot are a pair of vultures circling high in the sky. Perhaps they've spotted a new kill and are now awaiting their turn. Surely there can be no lack of carrion in a place like this. There are three species of saber-toothed cats alone here. The largest of them, *Machairodus*, is an extremely muscular, terrifying hunter the size of a lion. There are also hyenas around that could pose a danger to us as we walk. However, there is no sign of hyenas and saber-toothed cats, as both mostly stay hidden during the day.

Perhaps one of the chewed bones close to the ford is an example of the cats' work.

Unfortunately, we have no time to go back and take a closer look. The storm has finally arrived. Threatening clouds are gathering and darkening the sky. In a few minutes, a downpour will cover the land with a layer of mysterious red dust carried in on the rain. The dust is a crucial key to understanding this lost world, as we will see as we return to the twenty-first century.

The Secret of the Red Dust

Even the first people who dug at Pikermi near Athens about two hundred years ago and found fossils noticed that the bones were lying in a brick red, fine-grained material. They thought these sediments had been deposited by former lakes and rivers and called them *terra rossa*, red earth.[9] The fossils Bruno von Freyberg discovered in Pyrgos Vasilissis in 1944, which included the lower jawbone of *Graecopithecus*, were also embedded in this striking reddish, rocky material.

Could it be possible that this layer of rock consisted of hardened dust deposits and that all the bones from large mammals had been buried by dust? And how would all this dust have affected the environment in which *Graecopithecus* lived? We know the ancient Greek poet Homer wrote in *The Iliad* of "blood-dripping rain-drops,"[10] rain colored red by grains of dust. In Homer's day, people often thought "blood rains" were a warning from the gods that disaster was on its way. Today, however, we know that blood rains are caused by dust storms that originate in the Sahara.[11] These storms are still regular occurrences in the region around the

Mediterranean, where they deposit about 7 ounces of dust from the desert per 100 square feet per year (20 grams per square meter). This makes dust from the Sahara one of the most important components of the red-colored soil in this region.[12] Were conditions similar 7 million years ago?

Deep deposits of Saharan dust at Pikermi, Greece, with an embedded layer of gravel indicating a stream bed

I began to doubt that the red sediments in the Athens Basin were deposits from lakes and rivers when I visited the site of the dig at Pikermi in 2014. Closer inspection of the material of which the red rock was composed revealed that although the particles were fine, they were not fine enough to be lake sediments. Geologists in the field develop a good feel for the grain size in sediments. Here is the rule of thumb: If you cannot see the grain with a magnifying glass and the sediment feels gritty between your teeth when you put it in your mouth, you have silt with particles measuring between 6 and 60 micrometers. That is one-tenth the size of a speck of finely ground flour and a common dimension for dust from the desert. The red earth from Pikermi did indeed feel gritty between my teeth—and interestingly, it tasted salty, salty enough that I shook a little over my sandwich.

At first, I thought that perhaps spray from the nearby Aegean Sea had deposited salt on the surface of the soil. However, when I dug down deeper, I came across salty dust once again, this time containing gypsum and sodium chloride, which is common table salt. The ions these salts contained proved that this dust did not originate in the ocean. It must have formed on land, perhaps when a salty lake dried up.[13] Moreover, these dust particles were between 5 and 30 micrometers, a size that can easily be carried by the wind over long distances.[14]

That was not proof, however, that the salty dust at Pikermi originated in the Sahara. It was only when we examined the non-salty parts of the dust that we discovered they carried a distinctive geological signature. Almost all of them were exactly 600 million years old, a characteristic feature of rocks created when the Pan-African mountains were thrust up. The grains of dust, therefore, came from the remains of ancient mountains in North Africa. That gave us proof that the red sediments around Athens were formed from particles from the Sahara.[15]

The layers below the surface were not thin sediment deposits but significant layers up to 115 feet (35 meters) thick. When they were forming, storms were dumping about 5.5 pounds of desert dust per 100 square feet (250 grams per square meter) every year on southern Greece, more than ten times the amount carried in the winds over the Mediterranean region in modern times and comparable to the amounts being blown around the Sahal region in Africa today.[16] The place where El Graeco and all the other animals we met on our walk lived was therefore extremely dusty. Every spring, dust storms regularly appeared and unleashed blood rains on the landscape. In these epochs, the area was literally drowning in dust. *Graecopithecus* is, therefore, the first potential early hominin found in fossilized deposits of dust.

In comparison with *Graecopithecus*'s environment, the habitat of modern great apes contains very little dust. There is only one other highly developed primate alive today that can also cope with the amount of dust in *Graecopithecus*'s environment—us. But despite the dusty conditions, the world of *Graecopithecus* was not dry. That became clear when I examined the fossil soils around Athens. They formed about 7 million years ago as these sediment layers eroded and organic material decomposed. They were later buried under new layers of rock and conserved. These paleo-soils are valuable sources of information for reconstructing the climate in that epoch. For example, moisture in the soil dictated which new minerals formed and influenced their structure. By examining the minerals, it is possible to estimate how much rain fell. Up to 24 inches (600 liters per square meter) of rain fell per year back then in the Athens Basin, mostly in winter and spring. Today, the average annual rainfall is 15.75 inches (400 liters per square meter), so about 33 percent less. And the temperatures, with an average of 71.6 degrees Fahrenheit (22 degrees Centigrade), were 7.2 degrees Fahrenheit (4 degrees Centigrade) higher back then than they are in Greece today.[17]

Species-Rich Mediterranean Shrub Savannah

How do we know what the vegetation looked like when El Graeco was alive? Fossil plant remains such as leaves or tree trunks are not preserved in the dust layers, and we found little pollen.[18] However, phytoliths, glass-like inorganic structures found in plant cells, have survived in the sediments for a long time in large numbers. Using these, we could prove the existence of palms, cypresses, oaks, and plane trees. As we saw on our journey back in time, these trees did not form a continuous forest in the Athens Basin but grew in open, park-like stands.[19] Shrubs also grew well in this landscape. We found evidence of hollies, myrtle, and tamarisk. There were a particularly large number of species of grasses and herbaceous plants. The latter, for example, were particularly well represented with ground elder and many species of thistle. Grasses dominated the flora, mostly tall, clumping prairie-type grasses and short ground-covering grasses that today are typically found in tropical and subtropical savannahs.[20] The latter are rarely found in Europe anymore, and up until recently, scientists believed these tropical grasses had never been native to Europe. Our finds proved otherwise. If you put this all together, you get a picture of the vegetation in El Graeco's time: a semi-open landscape with islands of trees and shrubs set in wide expanses of herbaceous plants and grassy meadows. This mix of vegetation—you might call it Mediterranean shrub savannah—no longer exists anywhere in the world today.

This picture is consistent with evidence of frequent brush-fires in the area around Pikermi. We found fragments of charcoal in the red dust, ranging from microscopically small particles to fragments visible to the naked eye. They are proof that in the dry phases, fires swept through this landscape on a regular basis, another feature typical of savannahs.

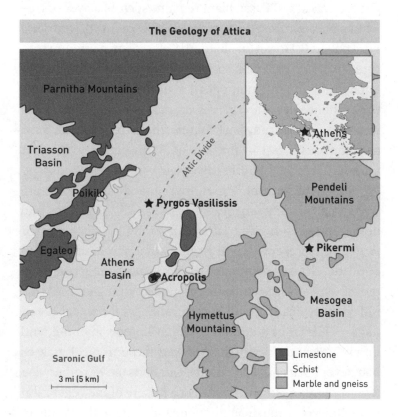

The animals we met on our journey back in time also belong in this landscape and are the ones for which the site at Pikermi has been famous for almost two hundred years. They are strikingly similar to the modern-day inhabitants of the African savannah. The resemblance is so strong that one of the first people to dig at Pikermi, Albert Gaudry, specifically compared the two.[21] The results of our research into the climate, vegetation, and role of fire in those times strongly support the conclusions he drew back in 1862. We can therefore rule out the possibility that *Graecopithecus frey-bergi* was a dedicated forest dweller as great apes are today. Forests require significantly higher rainfall to grow, especially in tropical and subtropical temperatures. And you do not get a significant buildup of dust in a forest environment. Savannah grasses, in contrast, are particularly efficient dust catchers.[22] The geology and the flora and fauna of the Athens Basin back then are sending a clear message: El Graeco, the oldest-known potential early hominin, settled a European savannah. For us, that is evidence that the traditional savannah hypothesis explaining the history of human evolution is still valid.

What Was El Graeco's Diet?

What did El Graeco eat in this environment? Could we perhaps reconstruct where in the Athens Basin he felt most at home? Access to drinking water must certainly have played an important role. Even today, water is the most precious resource in a savannah landscape. Especially in the dry season, many savannah animals in Africa undertake long and dangerous migrations to find fresh grass and the last remaining watering holes. The most famous examples are the fascinating migrations in the Serengeti. The situation in the Athens Basin at the time of *Graecopithecus* must have been similar. Even today, most of the water from the surrounding mountains gathers in a single small river, the Kifissos. It was built over last century and is no longer exposed to daylight, but this body of water, more of a stream than a river, existed back in *Graecopithecus*'s time. And Pyrgos Vasilissis, where *Graecopithecus* was found, lies only 1,650 feet (500 meters) to the west of where the Kifissos flows today. Above the layer of red rock where Bruno von Freyberg found the lower jawbone in 1944, there is a layer of coarse gravel that was left behind when the river changed its course.

We even found evidence of a nearby body of water in the stone itself. It was phytoliths once again that put us on the right track, in this case from sedges, rushes, and cattails—plants that need moist soil year-round. A stream course edged with sedges, rushes, and cattails, winding its way through a landscape dominated by grasses and herbaceous plants and dotted with stands of trees and shrubs, this was El Graeco's home. There he could always find water to drink and food. But what exactly did he eat?

If you look at the teeth in *Graecopithecus*'s lower jaw, you immediately notice how worn down they are. Not only on the chewing surface, but also in the spaces between the teeth. The paleoanthropologist Gustav Heinrich Ralph von Koenigswald wondered about this when he first examined the jawbone in Erlangen in 1969. In a scientific paper he wrote, he remarked that he had never come across anything like this in either a living or a fossil great ape. The spaces between the teeth had completely disappeared, and it looked as though one tooth was pressed up against the next. For anthropologists, however, this discovery is less surprising. Hunters and gatherers, especially those who live in savannahs and often chew a lot of tough fibrous plant matter, present the same patterns of wear on their teeth. As they chew, the teeth next to each other move up and down independently, just like keys on a piano. Over time, this causes one tooth to rub up against the tooth next to it. Although *Graecopithecus* was certainly not a very old male, the abrasion of his teeth was already very pronounced.

What did El Graeco chew on so regularly to put this much stress on his teeth? The tastiest food Pyrgos Vasilissis had to offer was the cattails. Over the centuries, humans have also valued these reed-like plants, which can grow up to 13 feet (4 meters) tall.[23] Almost every part of the plant is edible, tasty, and filling.[24] You can eat the shoots, stalks, leaves, flowers, pollen, or roots. The latter are rich in starch, and the shoots and the pollen contain protein, vitamins, and sugar—all of them essential nutriento. In Eastern Europe today, fresh cattail stalks are offered for sale as a delicacy at markets in the springtime and labeled Cossack asparagus or wild asparagus. Cattails have the added advantage that wherever they grow, they grow in large numbers. They are a food source that almost never runs out. But when eaten raw, the tough, fibrous shoots, stalks, and roots call for long and thorough chewing, as the lower jawbone of *Graecopithecus* attests.

However, El Graeco undoubtedly availed himself of other resources in his environment in addition to cattails. In the winter months, starchy acorns and the vitamin- and sugar-rich fruits of the strawberry tree were available. In spring and summer, in contrast, he would have varied his menu with the addition of vegetables such as sorrel, ground elder, chickweed, a brassica called wall whitlowgrass, dyer's weed, thistle, purslane, and sedge. We found evidence of these plants in the fossils at Pyrgos Vasilissis. All these wild vegetables were once commonly eaten, although today they have mostly been forgotten.[25] It is unlikely, though, that *Graecopithecus* stuck to an entirely vegan diet. Animal protein and animal fat from insects or bone marrow from dead animals must certainly have been widely available as well. We cannot definitively prove that he ate meat in any shape or form, but it is likely that he did. All in all, the range of food was similar to the food some researchers assume *Australopithecus* or early *Homo* probably, and in some cases definitely, ate.[26] At least as far as his eating habits were concerned, *Graecopithecus* was closer to modern humans than to modern great apes.

An artistic rendition by Velizar Simeonovski of *Danuvius guggenmosi* in its habitat

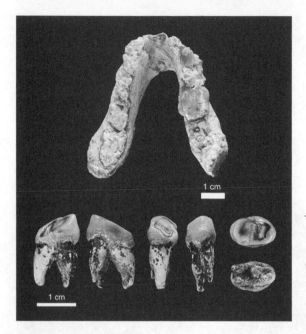

The lower jawbone of *Graecopithecus freybergi* from Pyrgos Vasilissis Amalias above photographs of the premolar from Azmaka taken from various angles

The happy discoverer of *Danuvius guggenmosi*'s lower jawbone, May 17, 2016

➔ Reconstruction of the facial skull belonging to Udo, a male example of *Danuvius guggenmosi*

⬇ Important bones from Udo, *Danuvius guggenmosi* (upper jaw, lower jaw, vertebrae, elbow, shin, thigh)

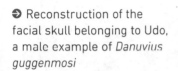

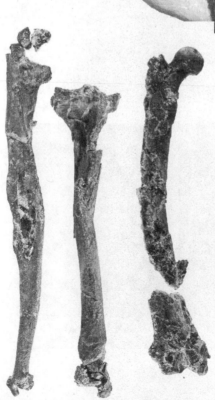

2 cm

The Foot of a Great Ape Compared With a Human Foot and a Footprint From Trachilos

Great ape

Modern human

Trachilos print

Analyzing the Footprint From Trachilos

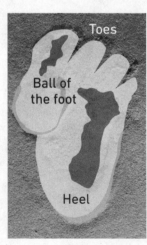

Toes

Ball of the foot

Heel

Footprint

Displaced sediment

Adhering sediment

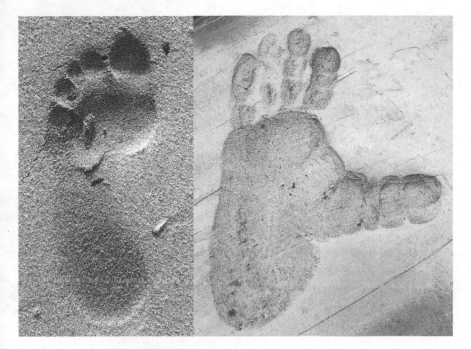

The left footprint from a human in sand compared to the left footprint of a chimpanzee in clay

The Grasping Foot of *Ardipithecus ramidus*

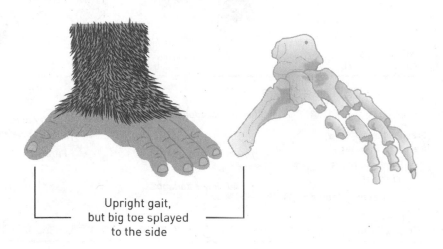

Upright gait,
but big toe splayed
to the side

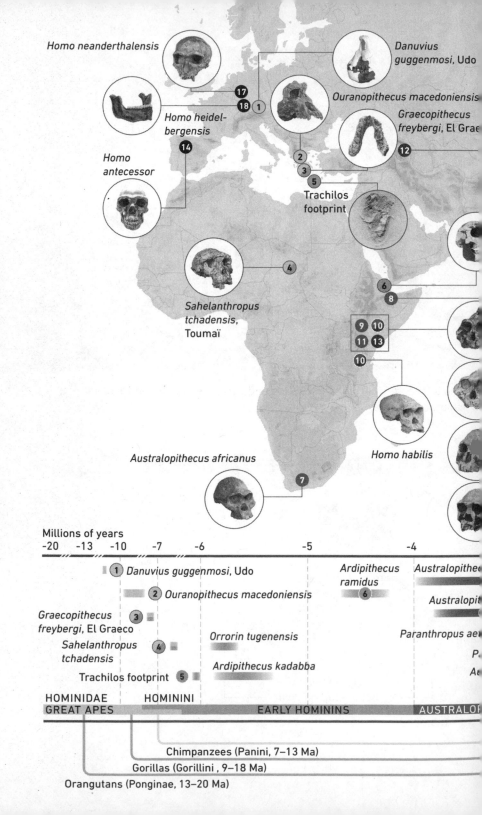

Homo neanderthalensis

Danuvius guggenmosi, Udo

Ouranopithecus macedoniensis

Graecopithecus freybergi, El Grae

Homo heidel-bergensis

Homo antecessor

Trachilos footprint

Sahelanthropus tchadensis, Toumaï

Homo habilis

Australopithecus africanus

Millions of years

| -20 | -13 | -10 | -7 | -6 | -5 | -4 |

(1) Danuvius guggenmosi, Udo

(2) Ouranopithecus macedoniensis

Graecopithecus freybergi, El Graeco (3)

Sahelanthropus tchadensis (4)

Orrorin tugenensis

Ardipithecus kadabba

Trachilos footprint (5)

Ardipithecus ramidus (6)

Australopithe

Australopit

Paranthropus ae

P.

A

HOMINIDAE   HOMININI

GREAT APES   EARLY HOMININS   AUSTRALO

Chimpanzees (Panini, 7–13 Ma)

Gorillas (Gorillini , 9–18 Ma)

Orangutans (Ponginae, 13–20 Ma)

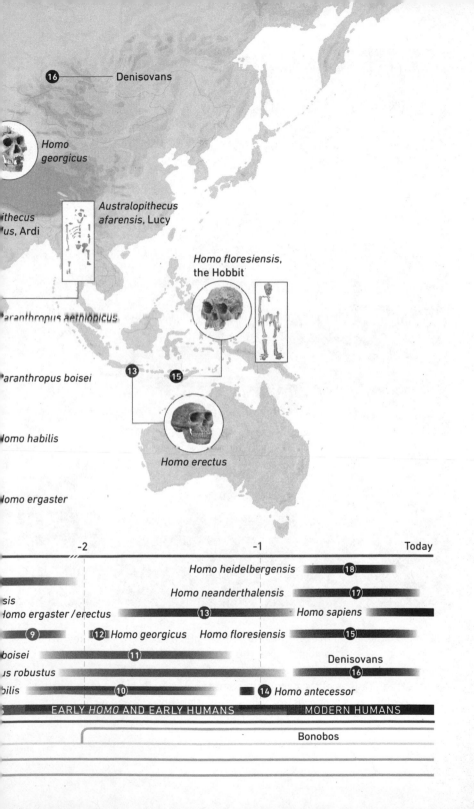

(16) —————— Denisovans

*Homo
georgicus*

*ithecus
us*, Ardi

*Australopithecus
afarensis*, Lucy

aranthropus aethiopicus

Homo floresiensis,
the Hobbit

aranthropus boisei

(13)   (15)

Iomo habilis

Homo erectus

Iomo ergaster

	-2	-1	Today
		Homo heidelbergensis (18)	
		Homo neanderthalensis (17)	
sis			
Iomo ergaster / erectus	(13)	*Homo sapiens*	
(9)	(12) *Homo georgicus* *Homo floresiensis*	(15)	
boisei	(11)	Denisovans	
us robustus		(16)	
bilis	(10)	(14) *Homo antecessor*	

EARLY *HOMO* AND EARLY HUMANS   MODERN HUMANS

Bonobos

A reconstruction of El Graeco's habitat. Although we do not really know what *Graecopithecus freybergi* looked like, the reconstruction of the landscape El Graeco inhabited is based on scientific data. Visible in the background are the limestone ranges of the Lycabettos and the Acropolis, along with giraffes (*Bohlinia*), elephants (*Anancus*), gazelles, *Hipparion*-horses, and a rhinoceros. Clouds heavy with dust from the Sahara darken the sky.

- 16 -

THE GREAT
BARRIER

A Gigantic Desert Becomes an
Insurmountable Obstacle

———

T HE RECONSTRUCTION OF the world in which *Grae-copithecus* lived proves that the story of the Sahara goes much further back than we have long supposed. But how did this enormous desert influence the evolutionary story of life so many millions of years ago? To get closer to an answer, it is worth taking a look at the Sahara and the people who live there today. This is the only way we can understand why deserts have always been hostile barriers of sand, dust, and rock that have left their mark on the distribution of plant and animal species far beyond their borders.

The Sahara is a tropical-subtropical hot desert where the sun is close to its highest point in the sky year-round. The intense sunlight heats the ground so much that in some places, temperatures near the ground can reach 140 degrees Fahrenheit (60 degrees Centigrade) or more. There is barely any precipitation. The central area of the desert has less than 0.4 inches (10 liters per square meter) of rain per year—one-hundredth the amount of rain that falls every year in New York City on average. The few plants that can survive under such extreme conditions get by on tiny amounts of water, such as the dew that collects overnight when the cool night air condenses. Every drop of water is valuable in this extremely dry region. It is hardly surprising that the only appreciable plant communities in the Sahara grow in oases or in the somewhat moister margins of the desert. Many areas in the heart of this desert are essentially bare of vegetation.

Of the relatively few animals in the Sahara, most are insects, spiders, scorpions, snakes, and lizards. Many species are crepuscular or nocturnal—that is to say, they are active at dusk and dawn or at night. To protect themselves from the heat of the day, they withdraw into underground hiding places or bury themselves in the sand. Mammals rarely live in deserts.[27] Those that do are usually small, like the other desert dwellers, and do not leave their hiding places until the sun sets. There is, however, an exception—an animal to which the rules of the desert seem not to apply and that is unique in the history of evolution: the camel.[28] This desert dweller is barely affected by hot and dry conditions because it has evolved to be perfectly adapted to this extreme habitat. No other animal even comes close. Camels are ungulates, hoofed mammals, but they have lost their hooves, because hooves would be useless in the desert. Instead, they have two extremely powerful toes padded on the undersides with large, round calluses, ideal for walking without sinking deep into the sand.

Camels have developed fascinating strategies to deal with the lack of water. They get most of the water they require from their food, so they do not need to drink much to supplement their intake. In addition, they have physiological adaptations to severely restrict the amount of water they lose in the hot desert climate. Their kidneys concentrate their urine and their gut extracts as much moisture as possible from their feces before they evacuate them. Their nose is constructed so that when a camel exhales, its breath condenses on its mucus membranes and the moisture is reabsorbed into its body. Last but not least, camels regulate their body temperature so that they store heat during the day and release it at night—a mechanism that ensures they rarely sweat. Equipped with these adaptations, camels can travel through the desert with ease and go for days without drinking. When they finally find water or are led to a trough, they are capable of downing large quantities of water extremely quickly. Their humps, incidentally, have nothing to do with water storage; they are where they store fat.

Despite these amazing adaptations, it was not until camels came into contact with humans that they became "ships of the desert," crossing these enormous arid regions to reach the farthest corners of deserts in Africa, Arabia, and Asia. This has been documented by genetic analysis, which proves that all the dromedaries (single-humped camels) in the world today are domesticated animals descended from a wild population that originated in the southeastern part of the Arabian Peninsula.[29] That was where humans first turned them into working animals about three thousand years ago, laying the foundation for a new nomadic lifestyle that eventually spread along the band of desert connecting the Old World from Asia, through Arabia, to North Africa.

In the course of these travels, the camel eventually arrived in the Sahara about two thousand years ago.[30] At the time of the Egyptian pharaohs, camels were not yet being used to carry goods or people. The Ancient Egyptian kingdom extended along the narrow, green Nile valley, where people lived in settlements and mostly used donkeys as beasts of burden, rarely venturing into the hostile expanses of the Sahara.[31] It was not until camels were introduced and a nomadic lifestyle developed, relatively late in the history of the region, that this enormous desert became somewhat more accessible, both biologically and culturally.

If we look even further back, we discover that the Old World deserts had an even more profound effect on the movement of animals at the time of El Graeco. They were barriers that could not be crossed. And that had consequences for El Graeco and his kin. All the *Graecopithecus* fossils found so far come from Pyrgos Vasilissis, near Athens, and Azmaka, in central Bulgaria. Admittedly, the Bulgarian finds, at 7.24 million years old, are about 60,000 years older than those in Pyrgos, but as we now know, that is only a short period of time in evolutionary terms. It is, therefore, unsurprising that along with *Graecopithecus*, we found the same species of animals at both sites. The situation at Pikermi, however, turned out on closer examination to be somewhat different. The Pikermi fossils, which we had dated as 7.3 million years old, were "only" 100,000 years older than the nearby fossils at Pyrgos, and yet the somewhat younger layers where *Graecopithecus* lay contained more species. What might explain the difference in species diversity between two similar sites?

Researchers have found many bones from extinct mammals that prove that in exactly this 100,000-year transitional period—between the finds at Pikermi and those at Pyrgos—numerous species of animals migrated to Europe. The most striking "new" contemporary of El Graeco's is *Anancus*, which is considered to be the first real elephant and is an ancestor of the elephants we have today. Researchers discovered ancestors of the elephant at Pikermi, as well, but these species were more primitive and still had four tusks—two on the upper jaw and two on the lower jaw. For this reason, scientists describe them as elephant-like rather than as elephants per se. *Anancus*, however, like modern elephants, had only two tusks protruding from its upper jaw. These were straight and could grow up to nearly 10 feet (3 meters) long. What is noteworthy here is that *Anancus* originally came from Asia, or more precisely, from the Indian subcontinent, and was a new arrival in Europe.[32] Could it be that El Graeco, too, was an immigrant and that he came to Europe along with other animals that were migrating to their new homes?

Early Climate Migrants

It was during this transitional period that the ancient Sahara and other Old World deserts reached the maximum extent of their first expansion. Multiple huge arid areas, stretching from North Africa, over the Arabian Peninsula, to the Gobi Desert in China, joined up to form an enormous dry belt about 6,000 miles (10,000 kilometers) long that split the Old World completely in two (see diagram on page 139). Such a mighty barrier could easily have driven whole communities of animals, including *Graecopithecus*, from their original

habitats to Europe. I, for one, think this hypothesis is plausible.[33] It is supported by the fact that this early expansion of
the Sahara at the time of *Graecopithecus* lasted a particularly
long time—700,000 years.[34] This is enough time to split the
lineage of a species that has been divided into two separate
subpopulations by a desert. Genetically isolated from each
other, they could evolve into different species. That could
explain why today, north and south of the Sahara, there are
African elephant shrews that are different from but closely
related to each other. Originally, they could have been part of
the same population.[35] This Saharan influence has also been
observed in freshwater fish and plants.[36]

But what happened when the first phase of Saharan
expansion ended? Today we know that over the past millions
of years, all of North Africa returned to green savannah when
the climate changed. You could think of the Old World desert
belt as a pulsating artery that contracted and then expanded
once again. The phases when the deserts expanded always
lasted longer than the green phases when they contracted.
Even today, therefore, an observer who is paying attention
can find the relics of these wet phases in many places across
the Sahara. Between mighty chains of dunes, there are flat
areas blasted by the wind. At first glance, they look like endless expanses of gray concrete. But if you look down, you
often notice amazing things: small bones with an unusual
pock-marked appearance, larger bone fragments, or even
complete turtle skeletons lying next to stone mortars and
arrowheads made of flint.

There is a reason for this. Where today endless desert
stretches out in all directions, there was once a large inland
sea. Just eight thousand years ago, soft-shelled turtles lived

here, along with air-breathing catfish with their distinctively shaped skulls. Around the inland sea were the fields and meadows of Stone Age farmers and livestock holders. In those days, the African monsoon blew its summer rains far to the north, and the amount of rain that fell was comparable to the amount that falls in London, UK, today. And so it should come as no surprise that early rock paintings found in the middle of the Sahara depict giraffes, elephants, and buffalo, savannah animals that today live a few thousand miles farther south.[37] The last wet phase in Africa began at the end of the last glacial period 14,000 years ago and ended 4,200 years ago.[38]

Looking further back, fossils show that in the green savannah phases there was an active exchange of animal species between Europe and Africa. For instance, 6.7 million years ago, hares that originated in Asia appeared in Africa for the first time.[39] They were among the first migrants when the early dry phase at the time of *Graecopithecus* ended. Also, 6.5-million-year-old fossil *Bohlinia* giraffes were discovered in Chad—and also in Pikermi.[40] Other migrants from Eurasia at that time included waterbucks, a species of large antelope that still live in Africa today, as well as the ancestors of goats and sheep.[41] In both Chad and Ethiopia, many of these new-to-Africa species were contemporaries of the potential early hominins *Sahelanthropus* and *Ardipithecus*.

This raises a question: Were *Sahelanthropus* and *Ardipithecus* also migrants from Eurasia, or had they evolved earlier in Africa and were now sharing their habitat with the new arrivals? The only way to answer this question definitively is to find more fossils. If we assume that our early ancestors did indeed migrate as the climate changed—looking for the best places to live and not letting themselves be constrained in their search by the boundaries that now exist between continents—the idea seems plausible. There is one more fascinating event that gripped the Mediterranean region 6 million years ago. It suggests that our early ancestors might have migrated more than once and that their opportunities to travel were not restricted to the African wet phases.

- 17 -

A GRAY-WHITE
DESERT
AND A SALTY SEA

The Mediterranean Dries Out

IMAGINE YOU ARE on the coast in southern Spain, walk--
ing from your hotel to the beach. Instead of encountering
cool breezes blowing off the blue Mediterranean, you are
confronted by an endless desert slashed by deep canyons.
The path down one of these canyons leads into a flickering,
glistening void, a moonscape where salt crystals, some as
large as you are, cover the ground. The abyss opening up in
front of you is at least 1.25 miles (2 kilometers) deep. Down
there, the temperature is over 122 degrees Fahrenheit (50
degrees Centigrade) and there is absolutely no sign of life.
This is not a description of a global catastrophe but a factual
account of the most extreme point of what is called the Messinian Salinity Crisis,[42] which occurred 5.6 million years ago
when the Mediterranean dried out.

Geologists detected the first traces of this extreme geological phase in the early 1970s. They were using the research vessel *Glomar Challenger* to drill deep into the bottom of the Mediterranean Sea when, completely unexpectedly, they came across massive layers of salt.[43] It was unexpected because salt usually dissolves in seawater and does not accumulate on the bottom of the sea. That is, unless the water evaporates so quickly that crystals form in the brine. Many researchers have been working on this discovery since then, but it has only been in the last ten years that they have developed a better idea of what exactly might have led to the salinity crisis.[44] So, what were the forces of nature that almost wiped out the Mediterranean at the end of the Miocene?

The Messinian Salinity Crisis

The Mediterranean region in the Miocene, ca. 7 million years ago

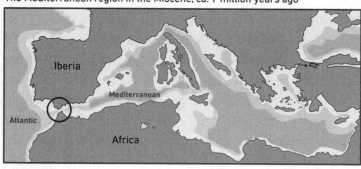

☐ Shallow water
☐ Deeper water
■ Deepest water

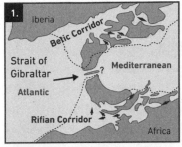
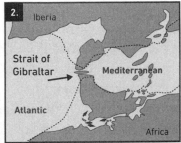

Dotted lines indicate modern coastlines. Arrows represent currents.

1. Late Tortonian ca. 7 million years ago
Two straits (corridors) connect the Atlantic and the Mediterranean.

2. Early Messinian ca. 6.3 million years ago
The Betic Corridor dries out. Water continues to flow through the Rifian Corridor.

3. Late Messinian ca. 5.6 million years ago
The Rifian Corridor also closes.

4. Zanclean ca. 5.3 million years ago
The Strait of Gibraltar is now the only connection between the Atlantic and the Mediterranean.

Isolated salty seas remain after the Mediterranean dries up, ca. 5.5 million years ago.

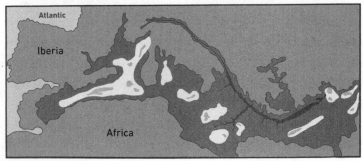

The crisis, it turns out, was triggered by tectonic forces in the Earth's core. For more than 100 million years the African continental plate had been drifting northward, pushing the seafloor between Europe and Africa under the Eurasian continental plate. The continents had been getting closer to each other and the sea between them had been shrinking. To the west, Africa and Europe got so close to each other in the Miocene that the connection between the Mediterranean and the Atlantic turned into a chokepoint. Today, between Spain and Morocco, we have the Strait of Gibraltar. Back then, there were two narrow channels: the Betic Corridor on the Spanish side and the Rifian Corridor on the Moroccan. Both channels were shallow, and their coasts were fringed with coral reefs. A large magma chamber filled with molten rock from the ocean floor lay under what is today Gibraltar, pushing the land above it into a convex dome. Volcanic eruptions were common events. When the climate cooled 6.3 million years ago, sea levels sank and the Betic Corridor dried up. Now the only connection that remained between the Mediterranean and the Atlantic was the somewhat deeper Rifian Corridor. As a result, the flow of water from the Atlantic into the Mediterranean was severely restricted—with drastic consequences.

Salty Brine on the Sea Floor

Even today, rivers and rainfall do not add enough water to the Mediterranean to make up for what is lost to evaporation. In the hot climate of the Miocene, the situation was even worse. The sea level began to drop and the concentration of salt in the water began to rise. After 300,000 years of this, conditions became so hostile to life that all the marine organisms in the Mediterranean died out. This was the phase when the unknown biped from Trachilos in Crete left its footprints in the sand. Did it have any idea that the sea was gradually retreating? Surely not, because at first the process happened very slowly, with the water receding at a rate of fractions of a millimeter per year. Perhaps the mysterious biped had been drawn to the coast by the large lagoons that were forming as the sea retreated. Here, it would have been able to harvest the last seafood offered by the Mediterranean before, finally, 5.97 million years ago, the salinity crisis set in.[45] At this time, the Riflau Corridor was perhaps 0.5 miles (1 kilometer) wide and 33 feet (10 meters) deep.[46] There was no longer enough space in the narrow channel for water from the Atlantic to flow in to replace water lost to evaporation, and that was the end of life in the Mediterranean Sea.

The inflow was sufficient, however, to slow the drop in sea level and, at the same time, it brought more salt[47] into the Mediterranean. As a consequence, concentrated brine accumulated on the sea floor. The water was so saturated with salt around this time that gypsum[48] precipitated into the water and was deposited in every part of the Mediterranean from Spain to Italy to Greece. This layer of gypsum in the geological profile of the Mediterranean is the oldest evidence of the salinity crisis. As the salt content continued to rise in the Mediterranean, the water was finally so saturated that rock salt began to be deposited over the layers of gypsum on the sea floor. Layer was stacked upon layer until the rock salt was an unbelievable 2 miles (3.5 kilometers) thick in some places. If you were to evaporate off all the water in the Mediterranean today, you would be left with a layer of salt no more than 100 feet (30 meters) thick. Over time, therefore, the amount of salt that entered the Mediterranean from the Atlantic via the Rifian Corridor increased a hundredfold.

These thick layers of salt were incredibly heavy, because salt weighs twice as much as water. They pushed the sea floor down by more than 2,600 feet (800 meters),[49] which caused the coastal areas around the Mediterranean to rise by about 60 feet (18 meters)—similar to the effect of pressing your fist into a bowlful of bread dough. This rise in elevation also occurred in the Rifian Corridor, causing it to finally dry out 5.6 million years ago. The Mediterranean was now completely cut off from the Atlantic and the salinity crisis was nearing its peak. In the hot climate, the sea level dropped by about 3 feet (1 meter) a year, even though large rivers such as the Rhone and the Nile were still contributing fresh water. Bit by bit, the Mediterranean was disappearing. Inland seas containing highly concentrated brine remained only in the deepest basins, 6,500 feet (2,000 meters) below current sea level. Finally, potassium salts were deposited on their floors. The Mediterranean was almost completely dried up.

What did the newly revealed landscape in the Mediterranean look like? In northern Italy, semidesert plants began to spread.[50] These drought-resistant grasses were among the few plants that could thrive on the margins of the Mediterranean. Because air temperature is affected by sea level, average temperatures on the salt flats during the summer likely hovered around 122 degrees Fahrenheit (50 degrees Centigrade). This hot air evaporated a lot of water. North and east of the deep basin in particular, the hot air rose, then cooled down, unleashing massive downpours over Central and Eastern Europe. The heavy rains, especially on the Balkan Peninsula, in combination with high temperatures, led to the kind of extreme weathering that happens in tropical climates. You can still see this in the dark red soil in this region, where iron ores formed as a result of this weathering process. Climate modeling also shows that the Messinian Salinity Crisis had global consequences, intensifying wind systems and increasing the intensity of storms, including rainstorms, around the world.[51] The latter often originated in the eastern Mediterranean and carried large quantities of dust and salt for thousands of miles. Today, for example, you can find 500-foot-thick (150-meter) layers of salt dust at the foot of the Zagros Mountains in Iran blown in from the dried-out Mediterranean.

Some of the rainfall was carried back down into the Mediterranean via rivers. These surged into the half-empty basin, carving deep trenches in its outer rim. Enormous canyons formed. Even though these chasms have since filled with sediment, you can still detect them via geophysical methods. These methods yield measurements that show that the canyons created by the Rhone and the Nile were up to 1.25 miles (2 kilometers) deep—deeper than the Grand Canyon. Even 560 miles (900 kilometers) upriver, near Aswan, the Nile carved out a trench over 2,500 feet (770 meters) deep.[52]

It must have been a breathtaking landscape. Erosion and salt as far as the eye could see—comparable to the views around the edge of the Dead Sea today. This Messinian landscape of salt and canyons lasted for less than 100,000 years—in geological terms, it was gone in the blink of an eye. The first changes came when the rivers from the north, which had been flowing into the dried-out Aegean, also cut deeply into the hinterland and finally tapped into the Black Sea, not far from what is today the Bosporus. Large quantities of less saline water from the Black Sea then overflowed into the hot, salty Mediterranean basin via the Aegean. This brackish water contained mussels and algae. Finally, some life was returning to the Mediterranean, and a huge inland sea formed: the Lago-Mare.

But even this lasted barely 100,000 years. To the west of the Mediterranean basin, near Gibraltar, erosion was busily reshaping the landscape. At first, there was just a small river with a waterfall that tumbled over a ledge, carving its course ever deeper as it flowed westward toward the Atlantic. Then, 5.33 million years ago, water flowing from the Mediterranean to the Atlantic succeeded in opening a channel and the Straits of Gibraltar were formed. Water from the Atlantic once again poured into the Mediterranean. The inflow, however, was not a catastrophic event, and it took 3,000 years for the Mediterranean to fill once again. After 667,000 years, the Messinian Salinity Crisis was finally over.

The Migration Story of Those Who Lived on the African Savannah

Shortly before the Messinian Salinity Crisis began, an unknown biped left its tracks in the sand on the Cretan Peninsula. What happened to it and the species to which it belonged? Did they survive the dramatic changes in the climate and the landscape? The extreme heat and the desert-like conditions? While we do not know for sure, we can assume that there was no way these early hominins could have survived close to the Mediterranean. Perhaps they migrated away, maybe even in the direction of Africa. But that is pure speculation. We do know, though, that the

Messinian Salinity Crisis triggered the migration of count-less other species of animals between Eurasia and Africa. For example, fossils of ancestors of today's camels and ostriches, two species that used to call the semidesert landscapes of Central Asia home, have been found by the Black Sea in the Ukraine.

Fossils of Asian immigrants have also been found in Bulgaria: a short-necked giraffe about the size of a bull moose called *Samotherium*. Far from the Mediterranean semi-desert, these creatures settled the woodlands with an as-yet unknown ape about the size of a baboon and the small European Old World monkey *Mesopithecus*, also found at Pikermi. African immigrants appeared in the western Mediterranean. When the strait between the Iberian Peninsula and Africa dried out, among the first animals to migrate to Europe were hippopotamuses and Nile crocodiles. Fossils of both these species were found at a site near Valencia, Spain, along with macaques and camels.[53]

Migration also changed the fauna in Africa. Camels[54] arrived in North Africa for the first time. Eurasian antelopes and carnivores such as bears and wolverines migrated to East African ecosystems and even South African ones. All are evidence of an intense faunal exchange in the Old World that had begun earlier but reached its peak during the Messinian Salinity Crisis. The reasons for these migrations were not only that newer and shorter migration routes had opened up between continents, but also that the boundaries and size of ecosystems were changing.

The sum of the evidence from contemporary research proves that the animal community of the African savannah arose from the Eurasian fauna found at Pikermi.[55] The origins of the savannah landscape, therefore, really do lie in Europe, in the place where El Graeco once lived. Nearly all the animals we consider representative of the modern African savannah originated in Eurasia: lions, hyenas, zebras,[56] rhinoceroses, giraffes, gazelles, and antelopes. The latter evolved into numerous new, purely African species after they arrived there. Others, such as lions and hyenas, remained species that lived in both Africa and Eurasia.[57] If the fauna of the African savannah[58] has its roots in Eurasia dating back more than 5 million years, why should early hominins be an exception to this rule? Is it not more likely that in stark contrast to the Out of Africa I theory, our early ancestors also crossed back and forth between continents? Indeed, fascinating finds from Asia prove that migration and the successful settlement of new ecosystems happened much earlier in our evolutionary history than previously thought. Are migration and curiosity perhaps deeply ingrained within us? Are they, perhaps, among the features that make humans human?

— Part 5 —

WHAT MAKES HUMANS HUMAN

FREE HANDS

Lots of Room for Creativity

TAKE A MOMENT to pay attention to your hands. It will be time well spent, because they are evolutionary marvels. Hold one up and examine it. Open and close it. Play with your fingers. Touch the tips of your four fingers with your thumb. Rotate your wrist. You should be able to turn it 180 degrees with ease. Ball your hand up into a fist until your thumb lies on top of and lends support to your index, middle, and ring fingers. That is something no ape can do.

Twenty-seven bones connected by joints and ligaments, thirty-three muscles, three main nerve branches, connective tissue, blood vessels, and skin equipped with highly sensitive touch receptors are behind the most delicate and most complex tool for grasping and touching that evolution has ever produced. The palm is protected by a massive sheet of fibrous tissue that makes it possible to grip things powerfully. The fingers are slender and small-boned, partly because they contain no muscles. They are controlled remotely, like puppets hanging from strings. But those strings are highly flexible, resilient tendons attached to muscles found not only in the palm of the hand and forearm but also all the way up to the shoulder.

Between this equipment and our complex brains, we can do things no other creatures on the planet are capable of doing: kindling fire, gathering the finest kernels of grain from the ground, knitting, cutting, knotting nets, turning tiny screws, typing on a keyboard, or playing basketball or a musical instrument.

Our thumbs have a special role to play in our dexterity. We can easily match them up with any finger. That allows us to feel and touch, to grab and hold. The saddle joint at the base of the thumb rotates like a ball joint. Our thumb is much longer, more powerful, and more flexible than that of our nearest relatives, the great apes. It allows us to execute a delicate pincer grip as easily as a powerful pinch. Chimpanzees can also clamp small objects between the sides of their thumbs and their fingers, but much less forcefully and without any sensory input from their fingertips. That means they have no means to hold or move tools such as pens or screws precisely between the tip of their thumb and their other fingers.[1]

A great ape holds larger tools—a stick, for example— pressed into their palm at right angles to their forearm. There are not many other options available to them. In contrast to chimpanzees and gorillas, we have highly flexible wrists that allow us to hold an object so that it becomes an extension to our forearm. This intensifies the force of a blow. It also means enemies and dangerous animals can be kept at arm's length. If an animal does come within range and full advantage is then taken of the extra leverage, bones can be broken.

It is not only the flexibility granted by the fully opposable thumb that makes the human hand so special, but also its extraordinary ability to feel and to touch. It operates almost like an independent sensory organ. We use it to feel the temperature of a breeze and of water. With its help we are able to fit a key directly into a lock, even in the dark. We can detect uneven surfaces with our fingers that we cannot see with our naked eye. With a little bit of practice, we can use our fingers to tell real silk from synthetic silk or real leather from fake leather, even with our eyes closed.

Our sense of touch detects delicate differences and sends this information via a dense network of receptors and neural pathways to our spinal cord and from there to our brain. Our fingers can even replace our eyes as ways to perceive the world, as the Dutch paleontologist Geerat Vermeij, who has been blind since the age of three, can attest. A specialist famous for his work on marine mussels and their ecosystems, he has never seen a fossil. Out in the field, he feels the complex morphological structures of mussels and of the rocks in which they are found. With his fingers, he "sees" details many sighted scientists miss. There is no doubt about it: our hands are an exceptional development in the history of evolution.

Sensory signals reach a specific region in the cerebral cortex, depending on the physical location of the sensory cells that detect them. The signals arrive in what is known as the somatosensory cortex, a flat band that stretches from one temple to the other. Every point on the surface of the human body is represented in the brain. Like every other body part covered in skin, our hands also have "copies" in the brain composed of neurons. Because neuronal pathways cross when they enter the brain, the left side of the body is represented on the right half of the brain, and the right side of the body is represented on the left half of the brain. Areas that are next to each other in the body are also next to each other in the brain. This arrangement begins at the top of the head, with zones for the toes and the soles of the feet, and ends at about the level of the temples, with zones for the lips, the tongue, and the throat.

The sizes of the zones, however, are not proportionate to the actual area occupied by the part of the body they represent. Body parts with a higher density of sensory cells, such as the fingers or the mouth, have a greater area of the brain devoted to them. That means the brain interprets the messages sent by these body parts more precisely, much like a camera gives a sharper picture when the pixel count is higher. Body parts that do not execute such finely tuned movements and are not as sensitive have relatively small areas of the cortex dedicated to them.

If you were to reconstruct a picture of the surface of a human starting with a map-like image of the cerebral cortex, the picture would be highly distorted, a so-called homunculus with a small, delicate-looking body and huge hands with enormous thumbs. The more sensors an area of skin has and the more intensively we use these sensors, the larger the area of the brain that represents it. People who regularly play the piano, for example, have larger areas of their brain devoted to sensing and moving their fingers.[2] If you were to sketch a homunculus for apes, the image would show considerably smaller hands and relatively tiny thumbs.

Our brain works with the anatomical possibilities latent in our hands and develops them into extraordinarily sensitive organs capable of extremely complex tasks. It is thanks to the interplay between the touch sensitivity in our fingers and the complex motor control coordinated by our brain—which co-opts the assistance of specific muscles in our arms, back, buttocks, and legs—that our hands can throw objects with such precision. Chimpanzees throw branches, dirt, and feces around. But even with a lot of practice, they would never be able to deliberately throw a basketball into a basket many yards away. The ability to use our hands to throw first stones and later spears with pinpoint accuracy became increasingly important over the course of human evolution. It was a necessary skill for our ancestors if they were to pursue successful careers as hunters.

The Origin of the Human Hand

How did a precision tool like the human hand, a tool that seems to have been at least as important for the process of becoming human as our upright gait, develop? The evolutionary ball started rolling, of course, when walking on two feet meant the hands were no longer needed for locomotion. Now they could be used more often for a wide range of tasks: transporting food or offspring, scooping up water, gathering material to build a shelter, or holding objects in one hand and manipulating them with the other to carry out specific tasks.

The more skilled our ancestors were with their hands, the more successful they were, and, therefore, the higher the survival rate of their offspring. And so, advantageous adaptations in hand structure prevailed as natural selection took its course. The evolution of our brain and our anatomy advanced in lockstep. The balance between hand bones, tendons, muscles, and nerves was constantly being refined, as were the hand's increasingly sensitive sense of touch and the brain's ever-more sophisticated oversight of motor coordination as it increased in volume and complexity.

We can trace the evolution of our hands back to the very beginning of the primate ancestral chart over 70 million years ago. The development of the primate hand probably started with small primate ancestors that originally lived on the ground and gradually conquered the tree canopy as their new home so they could help themselves to the abundance of fruits, buds, leaves, and insects available in the branches. Those that could grasp small objects clearly had the advantage.

The decisive course adjustment for the evolution of our hands happened when the human lineage and the chimpanzee lineage went their separate ways. The grasping hands of the earliest hominins were likely similar to Udo's.[3] They probably looked very different from the hands of modern chimpanzees, because the longer fingers and shorter thumbs of chimpanzees are a later adaptation to life in the trees.[4] Up in the canopy, great apes prefer to use the so-called hook grip, where the four fingers are bent into the shape of a hook and the thumbs are not involved. This makes it easier to swing through the branches. A longer thumb would get in the way.

For a long time, scientists thought that the early members of the genus *Homo* started out equipped with a hand anatomically similar to the hand of a modern human. This notion can be traced back to a few spectacular fossil finds in Africa from the early 1960s. There was great excitement in May 1964 when the primate researcher John Russell Napier, the paleoanthropologist Phillip Tobias, and Louis Leakey reported that over the course of many years of working in the Oldu-vai Gorge in Tanzania, they had found remains, including many hand bones, of the first humans to make tools. "The hand bones resemble those of *Homo sapiens sapiens*," they wrote, because from the individual fragments, they had reconstructed a hand that had especially powerful joints at the base of the fingers and a prominent thumb.[5] At the time, news of a humanlike hand that was 1.8 million years old caused a firestorm of interest.

The hand fragments were one of the main reasons the researchers attributed the bone finds to an early human, standing no more than 4 feet (1.2 meters) tall, that they called *Homo habilis* (Handy Man). That is controversial to this day, because a row of teeth found at the same time are a match for an early hominin of the genus *Australopithecus*. What is not in dispute is the special nature of the hand bones, which show clear evidence of a hand that was already strikingly human in appearance with a relatively long, quite flexible thumb. John Napier wrote in his book *Hands*: "The hand without a thumb is at worst, nothing but an animated fish-slice, and at best a pair of forceps whose points don't meet properly. Without the thumb, the hand is put back 60 million years in evolutionary terms to a stage when the thumb had no independent move-ment and was just another digit."[6]

Despite all the debate around *Homo habilis*, its relatively sophisticated hand shape was a good fit with the pebble tools of a similar age found in the Olduvai Gorge. Whether *Homo habilis* was a handy early human or a handy early hominin, there was no doubt that nearly 2 million years ago, the inhab-itants of Olduvai had taken a hammerstone in one hand and struck it against another stone to manufacture a stone tool with a sharp cutting edge. The brains of these gorge dwellers were approximately half the size of ours and the functional potential of their hands was not yet developed, but their hands were definitely no longer the hands of an ape.

Flexible hands and simple stone blades allowed the gorge dwellers to occupy a new ecological niche in the savannah-like landscape they called home: that of carrion eater. There were numerous ungulates and other large mammals grazing on the extensive grasslands, and they often fell victim to large cats. After the carnivores helped themselves, there was usually nutritious meat left over that could be quickly cut and scraped from the bones with sharp-edged stone tools—preferably before the hyenas or vultures arrived.

Two American archeologists, Kathy Schick and Nicholas Toth, did field tests in the East African savannah to see how well this would have worked. They tried cutting and scraping dozens of carcasses, including two elephants, using primitive stone tools. "We were amazed," they wrote, "as a small lava flake sliced through the steel gray skin, about one inch thick, exposing enormous quantities of rich, red elephant meat inside. After breaching this critical barrier, removing flesh proved to be reasonably simple, although the enormous bones and muscles of these animals have very tough, thick tendons and ligaments, another challenge met successfully by our stone tools."[7] When these primitive tools were wielded by modern humans, it was clearly a quick and easy job to use them to cut meat. Adding meat to the menu was a crucial step on the way to becoming human—up until then early hominins had likely mostly eaten plants. The increased protein intake must have led to better health overall and, in the long term, helped increase the size of the brain.

In the years since then, a great deal of evidence has surfaced that the first humanlike hand appeared considerably longer ago than 2 million years. For instance, in 2010, researchers in Ethiopia reported that they had found 3.4-million-year-old bones in Ethiopia with conspicuous notches that looked as though they had been made when meat was cut or scraped off them.[8] Other experts, however, interpreted the notches as bite marks.

One year later, scientists working on the west bank of Lake Turkana in East Africa found 3.3-million-year-old stones that they interpreted as showing signs of having been manufactured.[9] They are larger and more crudely fashioned than the stone tools from the Olduvai Gorge. The tools at Olduvai were more than a million years younger and could not be differentiated from stone fragments created by natural processes without reference to the geological and archeological context of the location where they were found, so it is difficult to say whether these considerably older objects really were tools.

Scientists at the Max Planck Institute for Evolutionary Anthropology in Leipzig, Germany, and the University of Kent in the UK have used newly developed methods to compare the internal structure of the hand bones of humans, chimpanzees, and early hominins. The spongy network of tissue inside bones is constantly being altered over the life of an individual in response to pressure. This means you can deduce with a relative degree of certainty the principal ways the hands moved. Their investigations showed that the inner structure of the thumbs and metacarpal bones of *Australopithecus africanus* is similar to that of modern humans.

From this, the scientists concluded this early species of hominin, which lived 4 million to 3 million years ago in southern Africa, could get a firm grip around an object with its thumb and the rest of its fingers and hold it tight—the kind of grip needed for tool use.[10] Jean-Jacques Hublin, from the Max Planck Institute in Leipzig, explained that the scientists believed, "There is growing evidence that the emergence of the genus *Homo* did not result from the emergence of entirely new behaviors but rather from the accentuation of traits already present in *Australopithecus*, including tool making and meat consumption."[11]

From Getting It to Gesturing It

An important aspect of human evolution is that hands are used not only for touching, working, throwing, or fighting, but also for communication. There is some evidence that the evolution of the hand had a significant influence on the development of speech. There is no direct evidence, of course, but you can deduce this indirectly by observing our closest relatives, the great apes, or by watching small children as they acquire language, using hand gestures to indicate what they want long before they say their first words.

For humans, gestures are an important component of expression. They both precede and accompany speech. They emphasize what is said and convey emotion. They can signal dismissal or acceptance. They can threaten, or they can express, elicit, and offer sympathy. In the sign language used by those who cannot hear, gestures almost completely replace words. Many scientists assume that gestures and sounds developed together over many millions of years to create increasingly complex forms of communication, mutually supporting and supplementing each other.[12]

Chimpanzees, bonobos, gorillas, and orangutans are also capable of communicating with gestures—although their repertoire is extremely limited. A field study carried out by British scientists in 2018[13] recorded more than two thousand separate observations and documented thirty-three different gestures. The vast majority were simple orders, such as "Give me that!" "Come closer!" "Groom my fur!" "I want sex!" or "Stop that!" All these gestures serve to start or stop a specific behavior. The researchers found that chimpanzees, gorillas, and orangutans not only used most of these gestures but also used them in the same way. What has not yet been explained is the extent to which these gestures need to be learned or whether they come naturally.

Michael Tomasello and his team from the Max Planck Institute in Leipzig have been searching for the origins of language for the past two decades. In numerous experiments in which they compared human behavior with the behavior of apes, they have discovered that human gestures went way beyond the simple orders given by apes. Apes indicate things that are useful to them at that moment. Human gestures often have a social context. They indicate things that might be of use to others or express emotions and attitudes that are relevant to the community.[14]

It seems it all started with gestures centered around self-interest and then, sometime in the story of becoming human—it is difficult to say exactly when—gestures were added to share experiences, intentions, interests, and rules. Tomasello is convinced that communication originated when early humans started pointing to things to show them to others.[15] For example, an early hominin would point to a vulture that was circling over a recently killed animal—a promise of nutritious meat—a place where nutritious roots were buried underground, a small child that had distanced themselves from the group as they went off to explore.

At first, pointing gestures had helped coordinate communal activities such as hunting or child minding. Later, they evolved into more complex signs for concepts, such as a fluttering movement to indicate a bird or cradling the arms to indicate a baby. According to Tomasello, sounds were then added to augment and expand this language of gestures. This corresponds with the American psycholinguist David McNeill's idea that gestures are basically nothing more than thoughts or mental images translated into movement. Having the hands free was therefore necessary for speech as we know it today to evolve in the first place.[16]

– 19 –

WANDERLUST

Curiosity About the Unknown

RE CURIOSITY AND the desire to settle new areas the characteristics that set the first modern humans apart? Or do they belong to the millions of years it took to make us? Only about one-fifth of the Earth's surface remains undisturbed enough today to be called something approaching "wilderness."[17] We have left our mark on the rest of it in one way or another. People have settled almost every corner of the Earth, even areas that are hard to reach, with the exception of Antarctica. And they did this earlier than previously thought.

For a long time, we believed that our ancestors reached remote islands far out in the ocean just a few tens of thousands of years ago at the earliest. Only the anatomically modern *Homo sapiens* with its big brain would have had the technology and intellect to be able to build boats and navigate the oceans. Or, at least, that is what we thought. We did not think earlier species of human would have been able to complete such a culturally complex task. At first, archeological finds seemed to confirm this assumption. The oldest boat in the world—a dugout canoe from Mesolithic times—is about ten thousand years old.[18] It was probably used only on small patches of open water in the fenlands and was certainly not seaworthy.

There has been a series of finds in Asia, however, that shattered the idea that *Homo sapiens* was the first seafarer and explorer in this world. Those finds suggest it is far more likely that the ability to cross oceans goes back more than a million years to the early members of the species *Homo*. The best-known of them was made on the Indonesian island of Flores in 2003. This early species of *Homo* became world famous under its nickname, the Hobbit. *Homo floresiensis*—its scientific name—was not only a hither-to-unknown species of human, but also one that did not fit with any of the prevailing theories of human evolution.[19] For me, that was reason enough to get a better picture of this puzzling hominin for myself, and so, in the spring of 2015, I traveled to Indonesia just as the rainy season was drawing to a close.

Visiting a Real-Life Hobbit

Flores, a volcanic island thrust up in one of the most tectonically active regions of the world, lies just 8 degrees south of the equator. Here, the Australian continental plate is being pushed under the southeasternmost tip of the Eurasian continental plate at a speed of about 2.5 inches (6 to 7 centimeters) per year, thrusting up a massive island chain in South Indonesia that stretches for many thousands of miles (see map on page 223). In addition to volcanic activity, there are also seasonal monsoon winds that affect the topography and vegetation on Flores.

The combination of volcanoes and weather leads to a mixture of very different types of landscapes. Thick tropical forests alternate with small open savannah areas where the only shade to be found is beneath a few scattered palms. These habitats on the west side of the island are where the 10-foot-long (3-meter) Komodo dragon lives to this day. A species of monitor lizard, it is the largest lizard on Earth and powerful enough to take down a buffalo. It is named after the small island of Komodo, which lies to the west of Flores, but its range is not restricted to that island. It is worth the journey just to meet one of these "living fossils."

The main attraction on Flores for me, however, was the Liang Bua Cave, the site where the Hobbit was found. The cave lies in the interior of the island and can be reached only by a car journey that is not for the faint of heart. The network of roads on the island is not well developed. On the few paved sections, pickups race past each other, focused on just one goal: delivering fresh fish from the coast as quickly as possible.

Everything changes, however, the moment you leave the main thoroughfares. The only option here is to drive slowly over bumpy dirt roads. This gave me plenty of time to observe the fascinating landscape of Flores. For the most part, the countryside has retained its original rural character, and only a few places have any tourism to speak of. What impressed me most were the symmetrical spider webs of lush green rice fields nestled in the hilly landscape, a sight you can see only on Flores. They are a testament to a long tradition of cultivating rice in a way that is unique to the island.

After we had been on the road for about three hours, the frequency of hairpin bends increased dramatically as the

narrow road wound its way for another 30 miles (50 kilometers) up a long chain of volcanic mountains, where the highest
peak reaches an elevation of nearly 8,000 feet (about 2,400
meters). These high ridges are remarkable for an island that
is little more than 3 million years old. Thanks to plate tectonics and volcanic activity, even today Flores is rising by about
0.02 inches (half a millimeter) a year.[20]

At the northernmost slope of the mountain chain, we
arrived at a heavily forested limestone massif. The unpaved
tracks in this area were still muddy from the recent rains.
About 100 yards (100 meters) down the slope, the primeval
forest river Wae Racang cuts into the rocky ground. There is
nothing in this landscape but small remote villages. Even at
a distance, the betel palms give them away, towering up over
the vegetation like cell-tower antennae.

After a journey that had lasted at least four hours, my traveling companions and I finally crossed a narrow metal bridge
over the Wae Racang. Three hundred yards (300 meters) farther on, we stopped in front of an unassuming wood cabin
with a sign that read "Museum Mini Liang Bua." It was not
clear to me whether this was a museum about a small person or whether it was referring to the size of the building.
There was no time, however, to sort this out, as the museum
director, Kornelius Jaman, was waiting for us at the entrance.
Jaman had participated in almost all the digs at the Liang Bua
Cave since 2001, and when no digs were underway, he gave
visitors tours of both the museum and the cave in which he
shared his passion for all that they had found.

On the slope directly behind the museum, barely 130 feet
(40 meters) above the Wae Racang, the 82-foot-high (25-
meter) mouth of the cave opens up majestically in the rock

face. Massive stalactites festoon the edges of the opening like the fringe on a tablecloth. The inside of the cave is even more impressive. Liang Bua is not part of a system of karst caves that lead deep back into the mountain. Barely 130 feet (40 meters) deep, it is a bell-shaped limestone cathedral flooded with bright green light from the primeval forest outside. Moss and algae grow on the stalactites, and the even dirt floor of the cave is a brownish color. When we were there, Jaman pointed to a rectangular depression in the ground, the very spot where, in 2003, LB1, the skeleton of *Homo floresiensis* was found 20 feet (6 meters) down. After the researchers stumbled across the fossils, it quickly became clear that nothing like them had ever been unearthed before. The find triggered a debate that continues to this day about where the Hobbit belongs in the story of human evolution.

To get a general picture of the debate and to answer the question about the significance of *Homo floresiensis*, let's first look at the anatomical peculiarities of the fossils and then review the unique paleo-environment of Flores at the time the Hobbit lived there.

The skeleton LB1 belongs to a female. She was about 3 feet 6 inches (1.06 meters) tall and weighed about 66 pounds (30 kilograms), which makes her about the same height and weight as Lucy in East Africa and Udo in the Allgäu. LB1 also shared her small brain size of about 400 cubic centimeters with early hominins and chimpanzees.[21] Other features of the skeleton indicate an early stage of evolution as well.[22] The pelvis and wrist are similar to Lucy's, and there is some evidence that the arch in the Hobbit's foot had not yet formed, another feature that points to a very early stage of human development.[23] Moreover, the bones on the top of the skull are considerably thicker than those of modern humans. The same could be said for the build of the shoulders, which look most like those of *Homo erectus*.[24] The Hobbit's arms are even longer than those of other early hominins and early humans, which makes it seem more apelike.

Homo floresiensis had especially large feet, which was rather fitting, given its nickname. The Hobbit's feet are, amazingly, 70 percent as long as its thigh. To put that in perspective, modern humans' feet are half as long as our thighs. A *Homo floresiensis* that stood 5 feet 7 inches (1.7 meters) tall would have had feet that were 12.6 inches (32 centimeters) long. That corresponds with a size 16 (European size 50) in shoes. People who wear shoes this large today are usually over 6 feet (2 meters) tall. With feet this size, *Homo floresiensis* must have had to lift its legs higher than we do when it walked, and it could certainly not run nearly as fast as modern humans. Yet, if you look at its teeth, tooth roots, and jaw shape, there are similarities once again with *Australopithecus* and other primitive early humans, such as those found at the world-renowned site of Dmanisi in Georgia that are still revealing their secrets today. All in all, the Hobbit appears to be a mosaic of numerous other finds from Africa and Eurasia.

The skeleton of LB1, *Homo floresiensis*, also known as the Hobbit

Because scientists could not explain this strange mixture of features, it was not long before someone proposed that *Homo floresiensis* was a modern human anatomically affected by disease. Those who supported this idea argued that numerous, often hereditary diseases lead to small stature, small brain size, and deformed bones in modern humans, and so surely some earlier human species had also been affected by these. The estimated age of the find initially supported this theory. At first, the scientists who discovered the Hobbit dated its age to eighteen thousand years ago, a time at which Flores had long been inhabited by modern humans. Also, along with the bones, the researchers had found stone tools and evidence of the use of fire, both cultural skills that are associated only with more highly developed forms of human—and which stood in contrast to the small brain size of *Homo floresiensis*. And finally, this opinion also fit with the assumption that primitive early humans would not have been able to cross the ocean to Flores, which remained an island even at the height of the last ice age when the sea level was 395 feet (120 meters) lower than it is today.

In 2016, however, it was discovered that the original dig team had overlooked an important geological marker in the cave deposits: a fifty-thousand-year-old area of erosion that angled through the deposits. It proved that the layer in which the skeleton was embedded had been freed at that time through natural processes and then re-covered with sediment. This process also explained why the stone tools and charcoal remains ended up at the same level as the bones of the Hobbit. The researchers took a closer look at LB1. They also analyzed the fossils of fourteen more Hobbits that had been unearthed during new digs in Liang Bua since 2003.

They found that all the fossils of *Homo floresiensis* from this cave were between 50,000 and 195,000 years old.[25] The Hobbit could not therefore be the remains of modern humans who had been stunted and misshapen by disease, because *Homo sapiens* first arrived on the island forty-six thousand years ago.[26]

There was a wealth of other important arguments against the idea that LB1 was a modern human deformed by disease. It is as good as impossible that in the span of the 150,000 years the species *Homo floresiensis* inhabited the cave it just so happened that only the bones of diseased individuals remained. In addition, in 2014, 50 miles (80 kilometers) away in the So'a Basin in Central Flores in a dig outside the confines of a cave, the skeletal remains of three more individuals that resembled *Homo floresiensis* were unearthed[27] and were dated back even further, to 700,000 years ago. These fossils were even smaller than those found in the Hobbit cave. It seems that these early humans were once 2 feet 7 inches (80 centimeters) tall. Researchers also found stone tools at the site that were more than 1 million years old.[28] That means Flores must have been settled by very small humans more than a million years ago.

These were the thoughts that were going through my head when Jaman finally led us into his museum after our tour of the cave. There, in a glass display case, lay the skeleton of LB1 in all its glory. Even though the skeleton on display was a copy (the original is in Jakarta), I still stood in awe when faced with this fascinating testament to human evolution. Jaman gave us a detailed description of unearthing each and every bone and of the problems the dig team faced. The bones buried in the wet floor of the cave were very damp, easily damaged, and soft as butter. The glues and rock hardeners they applied as the bones were unearthed were not intended for use on damp material and many of the bones were damaged. Photographs on display showed how the fossils were laid out to dry on newspaper on a bed in a hotel room.

Of Dwarf Elephants and Giant Rats

If the Hobbit was not a sickly undersized modern human, what could it be? Other finds from the cave provided helpful clues, and Jaman has the original bones from some of them safely stored in his museum. The researchers in Liang Dua also found the remains of elephants, birds, reptiles, and rats. What is astonishing is that all these fossils are either unusually small or especially large. Take elephants from the genus *Stegodon*, which were distributed across large areas of Asia at the time. The examples from Liang Bua were only 5 feet (1.5 meters) tall at the shoulder, tiny in comparison with conventional stegodons, which were about the size of African elephants. The birds, reptiles, and rats in the cave, however, were gigantic.

The researchers dug out the bones of a marabou stork that stood nearly 6 feet (2 meters) tall. Today marabou storks are no taller than 5 feet (1.5 meters). The extinct Flores marabou stork also had thicker and heavier bones, which means it would have been incapable of flight.[29] Even the bones of Komodo dragons excavated from the cave were up to 50 percent larger than the largest modern specimens. Three species of the fossil rats that were dug up were veritable giants. Apart from the Hobbit and the elephants, the rats were the only mammals found at the site. The largest was the Flores giant rat, which is still found on the island today and measures 30 inches (80 centimeters) from its nose to the tip of its tail.[30] People who are native to the island value them for their meat, just as their ancestors did. After the tour, Jaman invited me to his home close by, where I could taste them for myself.

So, the picture you get of Flores when the Hobbit lived there is somewhat hazy. Only a few vertebrates lived on the island and, compared with their closest relatives on the mainland of Asia, they were either very small, like the human and the elephant, or unusually large, like the marabou stork, the lizard, and the rat. How had this bizarre animal world arisen? The ocean around Flores is so deep that the only way animals could have reached the island is by sea. However, Flores lies in the middle of what is known as the Indonesian through-flow and is surrounded by the strongest ocean currents in the world. The Indonesian throughflow is one of the central components of the global climate system, a sort of gigantic heat pump that transports 530 million cubic feet (15 million cubic meters) of warm Pacific water per second past Flores and into the cooler Indian Ocean—that is, 4 billion gallons (15 billion liters). Yet, all the species of animals found on Flores must have crossed the ocean to get there. They originally came from the western part of Indonesia—the elephants, the stork, the rats—or from Australia—the monitor lizard.[31]

It was easiest for the birds, because they could fly to the island. The stegodons must have swum across. Smaller mammals such as rats could have arrived on driftwood, which often builds up after storms. After they arrived on the island, the animals must have adapted to their new, remote habitat. The larger species, such as the elephants, got smaller, which allowed them to survive with less food and at the same time sustain a large enough population to maintain genetic diversity. A smaller population of elephants of a normal size would likely have led to inbreeding, which over time would have meant the demise of the species.

In contrast, meat- or carrion-eating animals, such as the monitor lizard or the marabou stork, faced no competition for food on Flores. In other ecosystems, cats or hyenas could have fought them for prey or carcasses. On Flores, however, the marabou storks clearly did not even have to fly to find enough to eat. They grew bigger and gave up flying altogether. The lizards, too, must have felt right at home. All they needed to do was grow bigger than they already were and they could hunt even the dwarf elephants. Life was probably particularly easy for the adaptable, omnivorous rats. They found plenty of food both on the ground and up in the trees—ideal conditions under which to grow large, very large. Even given huge distances and extreme conditions, like the ones around Flores, it seems that the arrival of land animals to colonize remote islands was not uncommon.[32]

Hobbits on the High Seas?

People, however, are anything but accomplished swimmers and cannot fly. Add to that the fact that a remote island cannot be settled over the long term—in the case of the Hobbits for more than a million years—by a handful of individuals. If inbreeding is to be avoided, a certain population level must be maintained from the outset.[33] Every successful and long-lasting colonization of an island by people, therefore, is a noteworthy cultural act that requires conscious decision-making, organized activities, and a basic understanding of human biology. Which members of the group should make the journey? Experienced older people or younger people of a reproductive age? How many men and how many women are needed to settle the new environment successfully? How many tools and how much in the way of provisions do you pack when you are making a journey into the unknown? What would be simply a logistical challenge today—you only have to think of plans for colonizing the moon or Mars—would have been an almost inconceivable undertaking and would have required a complex language for meaningful communication. This, incidentally, is one reason why I believe we need to rethink our ideas about such early human cultures.

Certainly, when many of us think about how these ocean crossings might have been made, we conjure up images of technologies such as boats or rafts. But there are other ways and means of getting to islands, and there is evidence that they were used. In the 1980s, the Dutch paleontologist Paul Sondaar pointed out that elephants had a "snorkel" and that there might be a strong connection between the settlement of many islands by elephants and their settlement by people.[34] When an elephant swims, most of its body is submerged. Only the back of its head and, occasionally, a small part of its back is visible above the water, but with its trunk held straight up, it can still breathe. Elephants swim many dozens of miles in this manner. Copperplate engravings made by Dutch travelers in the eighteenth century show the native peoples of the Indonesian islands using swimming elephants as a means of transport to neighboring islands. Standing on the back of his elephant, the mahout takes the reins in his hands and steers his animal across the sea.

The paleontological investigations of Indonesia, in which Sondaar played a hugely important part, showed that over the past 2 million years almost every island was settled either by elephants or by their close relatives, the stegodons. The same holds true for the Mediterranean region.[35] Elephants and people were often the only large mammals that made it to these islands. That made the well-known South African paleoanthropologist Phillip Tobias suspect this had not happened by chance.[36] It is conceivable, for example, that early hominins or early humans noticed elephants purposefully walking down to the beach, swimming out past the horizon, disappearing from view, and never coming back. The animals were perhaps following scent signals or calls from their fellow elephants on remote islands that they were able to detect even over great distances.[37] And so it is possible that early humans sought out the animals and finally managed to cross oceans by riding on their backs. We know from the lifelong relationships that mahouts have with working elephants today that humans and elephants can form strong bonds of trust.

The Origin of the Hobbit

But how do these deliberations help us when it comes to the Hobbit? What if *Homo floresiensis* was not an island species, after all? Did it really arrive on Flores by chance? Was it small of stature because of limited resources in its new environment? It took a while for scientists to give serious consideration to such questions. Many researchers continued to champion the idea that the Hobbit shrank in size.

According to these scientists, the small *Homo floresiensis* was directly descended from the large *Homo erectus*, which lived on Java 1.5 million years ago.[38] A few of these early humans, they argued, somehow made it across the ocean to Flores and shrank, just as the stegodons had done. As evidence for this, they cited the lack of large prey animals on Flores. And, indeed, it is clear that the Hobbit mostly ate giant rats, as we can see from the remains of about two hundred of these rodents in Liang Bua Cave. I might mention here that it is unlikely that rats alone, even giant rats, would have been a sufficient food source for early people as large as *Homo erectus*. However, the researchers also found a few dwarf elephants in Liang Bua that had been butchered for their meat. They are evidence that once in a while even Hobbits could catch and kill a stegodon. Even in miniaturized form, these elephants weighed up to 880 pounds (400 kilograms). Such large quantities of meat could also have fed larger early humans, especially if you consider they could also steal eggs, gather seafood along the coast, and snack on insects.

However, as we have seen, there is little in the Hobbit's skeleton to suggest that what we are dealing with here is a miniaturized version of *Homo erectus*. The most recent ancestral study of *Homo floresiensis*, therefore, breaks with this idea and places the Hobbit close to the origin of the genus *Homo*.[39] The authors of the study think there are two possible scenarios. In the first one, Hobbits share a common ancestor with *Homo habilis*, a species that has been proven to have lived at least 1.75 million years ago in Africa. The second puts the Hobbit right at the beginning of the *Homo* lineage, 2.8 million to 2 million years ago, a phase of early *Homo* evolution for which we have very few diagnostically conclusive finds in

Africa. This second interpretation is in accord with finds of tools in India and China that have been dated to 2.6 million years ago, and with the almost 2.5-million-year-old "mystery ape" from the Yangtze River, *Homo wushanensis*. In this latter scenario, *Homo floresiensis* would not even have been an especially small hominin, because many early hominins and early humans stood no more than 3 feet (1 meter) tall. That is the case for Lucy, for example, and also for whomever left the footprints in Crete. Small stature, up to only 5 feet (1.5 meters), is a characteristic of all early hominins. It was only with *Homo erectus*, 1.7 million years ago, that humans grew to be more than 6 feet (1.7 meters) tall.[40]

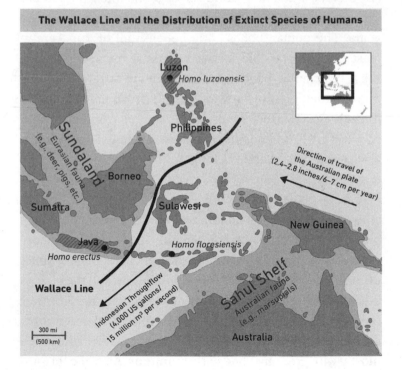

The Wallace Line and the Distribution of Extinct Species of Humans

A New Diminutive Human

The suggestion that the islands of Southeast Asia were settled by early humans recently received some additional support via a fascinating find. In 2007, researchers on Luzon, the northernmost main island in the Philippines, found a metatarsal bone from a human foot 9 feet (2.8 meters) below the current surface level of Callao Cave.[41] The scientists were researching the transition from hunter-gatherers to livestock holders in this area four thousand years ago. They were surprised when they found that the layers of earth in the cave were much older. They had dug down only 4 feet 3 inches (1.3 meters) when they came across stone tools, burned animal bones, and fire pits dating back twenty-six thousand years. The human bones lying beneath all of this were at least 67,000 years old. Buoyed by these unexpected finds, the researchers expanded their dig, and in 2011, they found more human bones—a row of teeth from an upper jaw, two finger bones, two foot bones, a fragment of a thighbone, and two teeth.[42] The finds belonged, as we now know, to two adults and one juvenile belonging to a new species: *Homo luzonensis*.

What is unusual about these fossils is that they do not fit with our current understanding of human anatomy and they present a completely new combination of primitive and more evolved features. The teeth from the upper jaw are very small, even smaller than those of *Homo floresiensis*, and the crowns are similar in shape to those of modern humans. However, the premolars from the upper jaw have three pointed roots, a primitive feature that up until now has been associated mostly with great apes and early hominins. Also, the forward molars are noticeably large in comparison with the back molars. This, too, is a feature that has been encountered before only in great apes, Udo from the Allgäu, and the African early hominin *Paranthropus* (Nutcracker Man).

The two finger bones from *Homo luzonensis* are especially interesting. They have a prominent curve to them, evidence that this mysterious creature was a good climber. This is another feature paleoanthropologists have seen only in great apes, as well as in early hominins and early members of the genus *Homo* in Africa. The finger joints of *Homo luzonensis* are constructed so that there was a limit to how far the fingers could be extended, something that had never been seen before. The two toe bones are also curved and show evidence of powerful tendons for curling the toes. Their size, shape, and structure, however, attest to an upright gait and resemble most closely the early hominin species *Australopithecus*.

In summary, according to the finds we have so far, *Homo luzonensis* is a two-legged human creature that could also climb well. At first glance, it might even be similar to the famous Lucy. However, because some of the features of the teeth look very modern, the species should be classified as belonging to the genus *Homo*.

But how could it be that creatures with such a mix of primitive and more developed features were living on Luzon millions of years after the disappearances of the African early hominins and early *Homo*? It turns out there is evidence that *Homo luzonensis* arrived much earlier in the island archipelago. In 2018, another research team published results from a dig in the northern part of Luzon.[43] Experts there came across the skeleton of an extinct Philippine rhinoceros, *Rhinocerorus philippensis*, that was at least 700,000 years old and had been completely dismembered. Three-quarters of the bones lay in an area of just 64.5 square feet (6 square meters). The ribs and metacarpal bones especially showed cut marks—a typical feature on bones when stone knives are used to strip away the meat and sinews. Both massive humerus bones were shattered, and you could see where they had been hit by blows, indicating that they had been smashed to reveal the fat- and protein-rich marrow inside. And as if any further proof were necessary, the scientists also found fifty-seven Paleolithic tools around the skeleton, including hammerstones and choppers along with the stone flakes that had been split from them. As there had been with the Hobbit, there was talk here in Luzon of a *Homo erectus* becoming smaller once it arrived on an island.[44] But once again, distinctive anatomical features of the fossils provided an argument against this.

Would it really have been possible for an early African *Homo* or perhaps even early hominin that could not walk on two feet for a sustained period of time to have successfully completed a journey halfway around the world, even though it was barely more than 3 feet (1 meter) tall and probably still spent a lot of time up in the trees? This scenario is not impossible. But perhaps these early hominins or early humans did not need to leave Africa, because they originated in Eurasia. Perhaps the evolution of the human family starts in the early grassland ecosystems of this enormous region and not in Africa after all.

Other researchers have also suggested that the discovery of *Homo floresiensis* and *Homo luzonensis* raises serious questions about the Out of Africa I theory.[45] This theory assumes that *Homo erectus* left Africa first, and no member of the human genus set foot on Asian or European soil before that. At the time of its alleged dispersal more than 2 million years ago, *Homo erectus* was already a long-legged, hairless early human, with a body shape very similar to ours and a relatively large brain. But its glorious departure as an upright human conquering new worlds, as imagined by so many, probably never happened.

There is one more important question in search of an answer here: Why would humans risk settling remote islands that they could not even see from the coast?[46] Perhaps wanderlust, the urge to travel into the unknown and conquer new places, is much older than originally thought and has been ingrained within us since our very beginning. You could also ask: Why do people today want to settle on Mars? And perhaps the answer to both questions is the same. We do this not because we can—after all, we cannot yet settle on Mars—but because we can *imagine* doing so, because we are mentally equipped to conjure up pictures of something we have never seen or done. We have the ability to construct mental images of worlds, places, and situations, and we can infuse these images with emotional and spiritual energy. The *idea* of traveling to whatever is on the other side, to the unknown, is perhaps a main driver of our evolution and this driver is in our heads. Our cognitive abilities are what allow us to dare to test our limits even when we do not have to, and, sometimes, to overcome biological and natural boundaries.

20

HAIRLESS MARATHONER

The Running Human

HEMERODROMES, OR "day-long runners," is the name given to the couriers in ancient Greece who could run for hours over long distances to deliver important messages. The most famous of these messengers was Pheidippides. In 490 BCE, the military commander Miltiades ordered him to run from Athens to Sparta to ask for help in the upcoming battle against the Persians at Marathon. If accounts are to be believed, Pheidippides ran the distance, 153 miles (246 kilometers), in under two days—a feat that seems well-nigh impossible. But why did the commander not send a messenger on horseback? The answer sounds unbelievable: because no horse would be capable of covering this distance so fast. Top athletes can run 60 miles (100 kilometers) in six or seven hours. The world record for a twenty-four-hour run is 188 miles (303 kilometers), and there is even a run that is 3,100 miles (almost 5,000 kilometers) long that the participants must complete in fifty-two days.[47] Many animals, however, can outrun humans over short distances. Predatory cats, horses, antelopes, wild dogs, and even kangaroos and hares, to mention just a few, can all run faster than the Jamaican runner Usain Bolt, the current world record-holder for the 100-meter sprint. Why is it that over short distances, a human is a slowpoke, but over long distances almost nothing can beat them?

One possible explanation lies in the hunting strategies of our ancestors. Even today, the San in Namibia and the Tarahumara in Mexico practice "persistence" hunting. The roots of this style of hunting likely lie far back in human prehistory.[48] In a persistence hunt, the prey, usually a gazelle or a deer, is chased by many hunters until it either stops because it is exhausted or collapses from overheating. The hunts usually last for hours and are often successful. Only wild dogs and wolves rival humans in this style of hunting, and so it is not surprising that in all likelihood, thirty thousand years ago or more, humans domesticated the first wolves and teamed up with them to hunt large ice-age animals such as mammoths and wisent.[49]

The origins of human persistence hunts, however, are certainly much older than this. The American anthropologist Daniel Lieberman has spent decades researching which anatomical and physiological features make humans so perfectly adapted for running, connecting the skills of the long-distance runner with our evolutionary story.[50] According to his hypothesis, the evolutionary story from early hominin to early human to *Homo sapiens* is one of continual optimization of performance running. What is important in this model is that upright gait developed first. Later, in the context of other independent evolutionary steps, our earliest relatives became endurance runners. How, then, does simply walking on two legs differ from running, or even racing?

The Art of Falling Forward Without Falling Over

The simple act of walking is a triumph of biomechanical engineering. When we walk, we move like an upside-down pendulum with our center of gravity above our legs. When we stand, we are balanced. When we walk, every step tips us slightly forward and to one side. We would fall over immediately if our knees did not compensate by alternately bending and straightening, allowing one leg and then the other to support the total weight of our body. We mostly depend on our feet to make sure this shifting of weight is converted into forward movement. Our feet roll forward from the heel to the toe and then our toes push our whole body forward. The soles of the feet are so small relative to everything they are supporting, even the simple act of walking requires skillful counterbalancing of the body, which is amazing when you stop to think about it.

Then you have running. When you run, every step propels you briefly into the air, and when you return to earth, your body cushions the impact and this action loads the ligaments and tendons in your feet with energy so they become like springs that are being stretched taut. As you take your next step, the ligaments and tendons relax, and the energy they release fuels your forward motion. There is also the fact that the body's center of gravity is in front of your feet when you run, making running like a continuous falling forward motion, where you catch yourself before you fall. Amazingly, at high speeds this means of locomotion is more energetically efficient than the pendulum swing of conventional walking, where the spring-loading of tendons and ligaments barely comes into play.

The human body has undergone many changes to perfect this exceptional form of locomotion we call running. Form and function are interconnected. The anatomical adaptations we have made have not only improved our overall stability but also our ability to regulate our body heat and store energy. All these features allow us to be superb runners if we choose to take advantage of them.

When we look at them up close, we can see how important these adaptations have been for human evolution. Let's start with our head. The development of upright gait by early hominins required the hole where the spine enters the skull, the foramen magnum, to move from the back of the neck to the bottom of the skull in the middle. This fundamental shift was necessary to make it possible for us to walk on two legs, while also balancing our head on our spine and orienting our eyes forward. This adaptation, however, was not sufficient for us to hold our head steady while running without expending a great deal of effort. And that was why humans also developed a powerful ligament at the back of our neck. It is attached to a bony ridge at the back of the skull and extends down to our lowest neck vertebra. Neither chimpanzees nor early hominins have this point of attachment, whereas in the early human *Homo erectus* and in Neanderthals it is even more pronounced than it is in *Homo sapiens*. When we run, this robust connection between our head and neck and our upper body is of critical importance because it means we can keep our balance as our legs move quickly.

At the same time, the physical disengagement of shoulders means they are free to swing along to the body's rhythm. Contrast this with modern great apes, which have strong muscles that attach their shoulder bones to their upper body. This mechanical connection is important for climbing, but it would impede the loose movement that is necessary for running.

In order to be able to angle our whole upper body forward and keep it balanced when we run, we have also developed large, powerful muscles in our rear ends. These act as a counterweight when our upper body leans slightly forward, as it typically does when we run. Our large gluteal muscle (the gluteus maximus) is the largest muscle in the human body, whereas in great apes, it is relatively small. There are also especially strong cords in the band of muscles that hold the back upright and allow it to rotate. These muscles connect the upper body to the pelvis, and you can see the bony protuberances on the skeleton where they attach to the coccyx and the pelvic girdle.

The anatomy of the human rib cage is also ideally suited for running long distances. Four-footed animals cushion the impact of running with their rib cages as well as their feet. As the animal runs, the rib cage contracts from side to side and then expands again. This means the frequency with which four-footed animals breathe is tied to the speed at which they move their legs, and they have to coordinate their gait and breathing. For each gait, therefore, there is a speed that is particularly energy-efficient for these animals. Thus, they can often only change speeds abruptly, like a horse does when it goes from a trot to a canter. Because people walk upright on two legs, there is no longer a close connection between the speed at which they move and the frequency with which they breathe, because the rib cage is no longer tied to locomotion. We can therefore seamlessly adjust the speed at which we run.

Let us now take a look at the lower half of the body, because the legs and feet of humans are also perfectly adapted to the requirements of endurance running. The knee joints are subject to special strain. Three to four times the weight of the body bears down on them with each step. And so it comes as no surprise that the surfaces of the joints between the thighbone and the shinbone are substantially enlarged so that the stress on the joint is better distributed. For regular walking on two legs, such an anatomical adaptation would not be necessary. That is why it did not exist in early hominins.

The length of the legs relative to body mass also affects the ability to run quickly and for a long time. This explains why, over the course of our evolution, our thighs became especially long to increase the length of our legs, and our upper body and arms became shorter to reduce the weight of our body. No ape or monkey has legs as long as a human's, not even the long-legged langurs, which can and often do take big leaps from one branch to another. Another important factor in our excellent ability to run is the well-developed, elastic arch in the foot of more highly developed humans, because it allows for much more powerful forward movement. For normal walking on two legs, a less highly developed arch, as we can see in the early hominin species *Australopithecus afarensis* and the relatively flat-footed Lucy, is sufficient. Because there is enormous pressure on the foot when we run, the metatarsal bones in humans are also closely aligned.

A foot adapted for running first appears in the history of human evolution in early members of the genus *Homo*. The arch and closely aligned toes are not the only features that make us better runners. Short side toes allow us to roll our feet from heel to toe efficiently. Longer toes would increase leverage and therefore increase the forward motion, but this would come at the cost of stability.[51] Clearly, on this point, evolution decided to err on the safe side, partly because if you fall when you're running at top speed, there's a good chance you will be injured.

And now let's turn our attention to our thickest and strongest tendon: the Achilles. It also has a crucial role to play in supporting us when we run. Because we have a shorter heel than either great apes or early hominins, the human Achilles is under more pressure and conveys more kinetic energy, similar to the taut string on a bow. Modern humans, *Homo sapiens*, have a heel bone that is even shorter than that of our extinct cousin the Neanderthal—evidence that from the very beginning we were more effective endurance runners than they were.[52]

It is not only the physical equipment we use to move that has been modified to make us better runners, but also our sense of balance. The inner ear helps us keep our balance, and if you look at the bony part of the inner ear, you can see that the semicircular canals—the tubes that register important data such as our orientation in space and how fast we are moving—are much wider than those of great apes and early hominins. This helps us keep our balance better when we are running fast.

The Staying Power of the Persistence Hunter

What about adaptations that leave no trace in bones or fossil footprints? How can we follow their evolutionary story? All we have to work with here is the comparison with ourselves and the assumption that even in the past, a nearly identical body structure would include similar physiological features. For example, we do not know exactly how the eyes of our early relatives were designed, but we do know that the muscles around our own eyes allow for rapid eye movements and that we register sharp images even at fast "shutter speeds," for example, when we are running.

Our sense of balance also allows us to get a uniform, detailed picture of our surroundings while we are moving fast. Vision, our awareness of our own bodies, and the organs in our ears that control balance are all closely connected. This ensures that our ongoing video stream of the world is not fuzzy and does not wobble like an amateurish home production. The video stays in focus and is seamless even when we are jogging, shaking our heads, or traveling in an off-road vehicle over bumpy terrain. The vestibular-ocular reflex is the mechanism that stabilizes the image. It is coordinated by the brainstem, where information from the semicircular canals in our ears is linked to muscle movements in our eyes. For every movement made by the head or body, the reflex initiates a precise countermovement in the eyes. It takes less than 8 milliseconds to react, making it the fastest reflex in our central nervous system.[53] You need a feature like this for good spatial awareness when you are moving fast, and it is likely that our early relatives profited from this evolutionary adaptation when they went out on a hunt.

The same can be said for our exceptional ability to protect our bodies from overheating. Modern humans possess one of the most effective inbuilt cooling systems evolution has ever produced. There are over 4 million sweat glands distributed over the human body. They ensure that when we work hard and it is hot, our body secretes pints of fluid that evaporate and cool us down from the outside. Many other mammals can sweat; however, compared to humans, they often have very few sweat glands. They also have fur that keeps the heat in. To cool down effectively, they need to pant, get rid of excess heat through overly large ears, wallow in mud, or take a bath. That is why it is often life-threatening for animals to cover long distances on foot, something humans have been taking advantage of in millions of years of persistence hunting. As long as we are drinking enough to replace the fluids and minerals we lose when we sweat, we can run for hours, even on hot days.

We can also run for a long time because we store energy in a different way from most animals. When animals need a short, sharp burst of energy, they mostly use glycogen, a form of sugar stored in the liver and muscles. That allows them to evade an attacking predator in a fraction of a second. However, an animal that gets its energy only from stored glycogen tires quickly. Those that want to run for long distances need to have fat reserves to tap into as well. This is not an option for many animals, especially those that live in warm climates. Humans, in contrast, profit to this day from a gene mutation that appeared in European great apes 15 million years ago and has been passed down along the evolutionary line to us. As explained earlier, thanks to the increase in uric acid in our bloodstream that comes with this mutation, we have no trouble converting sugar into body fat.

Lacking fur, anatomically equipped to perfection for endurance running and with one of the best cooling mechanisms any mammal has ever had, and physiologically energy-efficient to the highest degree—that is one way of describing humans as endurance runners. All these features make humans ideally suited to be persistence hunters. And once they are trained in this style of hunting, they can outlast almost any animal.

If we now take another look at the fossils of our early relatives, we can see that the first members of the genus *Homo* developed these traits long after early hominins had already stood up on two feet and become bipeds. And so this adaptation is an important chapter in our particular evolutionary story. The first species of human that we know for certain conquered the art of endurance running is the 1.8-million-year-old *Homo georgicus* from Dmanisi in the Caucasus.

21

FIRE, INTELLECT, AND SMALL TEETH

How Diet Influenced the Development of the Brain

───────

THREATENING STORMS BLOW over the brush savannah through the night. The wind howls. Thunder roars. Flashes of lightning light up the night sky. On a small hill, a dozen early humans cower together under a rocky overhang. They are tense and nervous. Shortly before dawn, there is a loud crack as lightning strikes an umbrella acacia a few hundred yards away. The tree immediately catches fire. Burning branches fall, igniting a localized brush fire. Their eyes wide, the early humans watch the blaze. The day is well advanced by the time the flames die down. Four males abruptly grab their handaxes and stone knives and set off to investigate.

They know from previous fires that they are likely to find small, but sometimes also large, animals that were caught in the flames. This charred meat is highly prized. It is much easier to chew and tastes better than strips of raw flesh. Today the scouting party does not have to search for long. A falling branch has landed on top of an antelope and killed it. Parts of the tree lie around the dead animal, some of them still burning or smoldering. The men quickly set to, using their tools to skin the animal and cut some of the meat off the bones. Suddenly, directly behind them, they hear the cackling calls of hyenas. Without even thinking about it, one of the men grabs a burning branch to drive off the dangerous competition.

It is only when the hyenas flee that the man realizes he is holding fire in his hand. The wood is hot and burning at one end, but he can hold onto the other end without hurting himself. Curious, but also respectfully cautious, the others want to touch the burning branch. They realize that if they are careful, they can carry the flames. The most important trophy they take home to the members of the group back at the camp today is not meat but fire.

That is one possible scenario. Discovering the usefulness of fire was even more consequential for the development of early humans than the invention of stone tools. Charles Darwin thought it was "probably the greatest ever [discovery] made by man, excepting language."[54]

News of the discovery would have spread. Fire pits were built, kept supplied with combustible material, tended, and guarded. They were not allowed to go out.[55] It took many generations before humans discovered how to light them.[56] If early humans had to leave a camp, they placed an ember from their fire in tree bark, covered it with leaves to protect it, and carried it with them. Fire was their most precious possession. It delivered light and warmth on cold, dark nights. It protected the group from prowling predatory cats and other animals. It allowed them to cook their food and provided them with security and comfort.

The First Traces of Cooking Fires

When and where humans succeeded in controlling fire for the first time is the subject of bitter academic debate, and we will probably never know for sure. The problem is that when you examine ancient traces of fire you cannot always know for sure whether the fire occurred naturally or was human-made. One unambiguous exception to this rule was found inside the Wonderwerk Cave[57] in the northern part of South Africa near the border with Botswana. An international team of scientists discovered in the interior of the cave—about 100 feet (30 meters) from the entrance and 6 feet 6 inches (2 meters) beneath the surface—numerous remnants of slightly scorched bones and ashes from plants right next to handaxes and other early tools that could have belonged to the early human species *Homo ergaster*.

From the location of the finds as well as the form of the ashy remains, the archeologists deduced that what they had found in the cave were not the remains of a natural forest fire but the remains of a fire that had been lit in the cave intentionally. The thickness of the ash layers also pointed to repeated fires at the same location. The researchers dated the traces of fire back at least 1 million years. When they examined the finds using a special microscopic technique—spectroscopy—they discovered that the pieces of bone had been heated to as much as 932 degrees Fahrenheit (500 degrees Centigrade). A wood fire normally reaches 1,472 degrees Fahrenheit (800 degrees Centigrade). In addition, there were no charcoal remains.

The scientists concluded that instead of using larger pieces of wood as fuel, the cave dwellers had used smaller pieces of plants, such as leaves, twigs, and dry grass. Because this plant material burns completely, all that is left behind is a fine ash that is easily swept away. This might also be a reason so little evidence of fire pits from the Later Stone Age has been found.

Less convincing finds of what might be fire pits in Kenya, in Ethiopia, and in Swartkrans, near Johannesburg, have been dated back 1.5 million years. In China, researchers even found 1.7-million-year-old charred bones from mammals near fossils of *Homo erectus*. Looking at the evidence through the lens of evolutionary biology, the Harvard University primate researcher Richard Wrangham thinks the use of fire must have played a central role in the further development of humans before or around the time this species appeared, which would be about 2 million years ago. If that had not been the case, he argues, these early humans would never have survived and would not have been able to persist for almost 2 million years in a wide variety of environments in Asia, Europe, and Africa.[58]

Anatomically, *Homo erectus* was very similar to modern humans. They were well over 5 feet (1.5 meters) tall, walked liked we do, had similar feet, and, in contrast to early hominins and the earliest members of the genus *Homo*, were certainly no longer any good at climbing trees. They lacked the features in their hands, arms, shoulders, and rear ends that would allow them to do this with ease. Therefore, they probably slept mostly on the ground, where they would have been exposed to carnivores and other dangerous animals such as rhinoceroses, elephants, or buffaloes, which could easily have lumbered through their camps. All they had to defend themselves against these threats was cunning, team-work, sticks, and stones—and perhaps fire, which warms, brings light into the darkness, and scares away predators.

From Eater of Raw Meat to Cook

Wrangham's key argument, however, hinges on the relatively small size of *Homo erectus*'s teeth. Between the time when *Homo habilis* was around and the time *Homo erectus* appeared, tooth size reduced more significantly than it ever had in the previous 6 million years of human evolution. Smaller teeth are a significant disadvantage for a creature whose diet consists overwhelmingly of tough raw meat, fibrous roots, and fruits protected by a hard rind. Wrangham is therefore convinced that the only adequate explanation for the small teeth of *Homo erectus* is that they had the ability to cook food over a fire or perhaps in hot springs. Food that has been heated is much easier to chew. According to Wrangham, the use of fire, and the use of tools to pound and chop raw food, meant that a powerful chewing mechanism was no longer necessary and, over the course of evolution, it was replaced with a less cumbersome mechanism.

Food that has been heated is also quicker and easier to digest and has more nutritional value than raw, and the amount of energy a person can extract from it increases enormously. Starchy foods such as grain or potatoes, for example, provide 30 percent to 50 percent more energy when they are cooked,[59] and cooked eggs provide 40 percent more useable protein than raw.[60] With foods that were easier to digest and the ability to extract nutrients from them more efficiently, humans developed smaller stomachs and digestive tracts, which in turn saved a significant amount of energy. This energy could be redirected to meet the needs of an increasingly energy-hungry brain. Roasting, grilling, and boiling also rendered a number of previously inedible foods fit for human consumption—such as the seeds from grasses or certain starchy tubers. Heat opens up hard, fibrous parts of plants. It also breaks down certain toxins in plants, kills pathogens and parasites, and helps conserve food.

Early hominins were mainly vegetarian. When early humans came along, they added raw meat to their diets. Wrangham, unlike some scientists, believes the key to human evolutionary success was not consuming raw meat, but exposing meat and plant foods to heat. He assumes that, as with the hunter-gatherer societies that still exist today, more than half the items on the menu for early humans were starchy roots, nuts, seeds, and fruits. "The reduction in tooth size," Wrangham wrote, "the signs of increased energy availability in larger brains and bodies, the indication of smaller guts, and the ability to exploit new kinds of habitat all support the idea that cooking was responsible for the evolution of *Homo erectus*."[61] And, if you look further forward toward the present, cooking was also responsible for the development of the human brain in its present size.

Intelligence Through Starch

The large, complex human brain is so hungry for energy that it uses up more than 20 percent of the body's energy needs and 60 percent of the glucose dissolved in the blood, even though it makes up only about 2 percent of the body's weight. An organism can afford such an extravagant organ only when it has enough fuel at its disposal at all times. It is highly likely that our fire-using ancestors got this fuel primarily from cooked vegetables with a high starch content, rather than from meat. Starch is made up of long chains of glucose and is one of the best sources in nature for this sugary fuel. These glucose chains resist digestion when intact; however, when starch is heated and the chains are broken down, we can digest almost all the glucose it contains.

Investigations made by my research team showed that, like us, our earliest ancestors were sugar junkies. We found fairly advanced dental caries in the 12.5-million-year-old teeth of a species of great ape called *Dryopithecus carinthiacus*.[62] The teeth were dug up in Kärnten, Austria, in 1953. We were extremely surprised by what we found. Up until then caries had always been connected with the invention of agriculture—the Neolithic Revolution—about ten thousand years ago because it was thought that this was when large amounts of cooked starch became a regular part of the human diet. Interestingly enough, modern great apes do not have a problem with caries. A comprehensive comparative analysis of the status of the teeth of 365 chimpanzees from the wild in West Africa shows that only 0.17 percent of their teeth had caries.

In a recent study, an international research team led by Karen Hardy from the Autonomous University in Barcelona also found extensive physiological, genetic, anthropological, and archeological evidence that the role played by starch especially, but also by other carbohydrates, was important much earlier than the academic world had suspected. According to Hardy, starch made a significant contribution to the astoundingly rapid development of the human brain.[63]

You can still read this story in our genes. The enzyme amylase, which is responsible for converting the glucose chains in starch into digestible simple sugars, is more prevalent in human DNA than in other primates' DNA. Recent genetic studies suggest that this particular characteristic developed about 1 million years ago. Hardy and her colleagues are convinced that using fire to prepare food and the increase in amylase genes co-evolved over a long period of time. The interdisciplinary team explained that although protein rich meat undoubtedly also played a role in the brain's evolutionary leap forward, cooked starchy food was what allowed us to become really smart. So, no intellect without fire?

Had they eaten nothing but raw food, our human ancestors would never have been able to maintain a high-performance brain that grew to be an average size of 1,300 cubic centimeters with over 80 billion neurons over the course of evolution. If they had lived off raw meat and plant material, our brain would never have developed. This is the conclusion Brazilian researchers came to after they took a close look at the feeding behavior of great apes alive today and assessed the energy requirements of their brains.[64] According to their research, if a mature gorilla that eats mostly leaves, flowers, and fruits had a brain that was proportionate in terms of body size to ours, it would have to eat for an extra two hours a day to fuel that brain. As gorillas already spend up to eight hours a day eating and digesting, this is almost impossible. There are simply not enough hours in the day for them to have bigger brains.

Chimpanzees also eat mostly leaves and wild fruit, and very occasionally animals they kill, such as colobus monkeys, and devote many hours a day to finding food. The food they usually find is so tough and hard to digest that it is already dark by the time they are finished chewing and digesting it. Stone Age humans would barely have been able to cover their energy needs if they had lived like that. They would have been gathering, hunting, chopping, chewing, and digesting from dawn until dusk. Even if they had eaten nothing but raw meat, things would not have looked much better. There would have been barely any time left over for making tools or for socializing. Thanks to cooked meals, however, humans spend only one-fifth to one-tenth of the time great apes spend eating and digesting. This allowed the fire tamers more free time for creativity and, perhaps, for gossiping around the campfire.

Along the evolutionary path to modern humans, fire paved the way for far more radical changes than just better food and more free time. Fire was and is dangerous, but it is something we cannot do without. Humans hardened the tips of their spears and arrows in the flames of their fires. They used it to bake molded, damp clay into ceramics. They heated rocks to extract the metal they needed to forge their weapons. They used fire to clear space in woodlands for their fields.

Step by step, with fire's help, humans made themselves less dependent on their environment. Eventually, the flames fed the development of human civilization. Today, we use fire in power plants and in the combustion engines of our cars, ships, and airplanes. We use copious quantities of fuel whose origins predate the dinosaurs. Almost 2 million years have passed since we first tamed fire, and today, about sixty-five thousand human generations later, we need to understand how profoundly our mastery of fire has changed not only us but also our environment, our climate, and our planet.

22

VOCAL CONNECTIONS

From Alarm Cries to Culture

===========

THE FUNDAMENTAL BASIS for human civilization is our ability to express our thoughts and feelings through facial expressions, gestures, and sounds. Without complex speech, cooperative tasks involving a lot of people would be impossible. If this evolutionary wonder had never happened, there would be no agriculture, no trade, no religion, no nation states, no literature, and no art. Nothing advanced the development of humans as quickly as spoken, and later written, language.

Language allows us to communicate with other people and give them a glimpse into our innermost thoughts, our motivations, our moods, and our points of view. Language unites us and saves us from being alone. Language helps us think. It helps us exchange ideas with other people, develop problem-solving skills, and expand our knowledge. In other words, it helps us survive. It is the key to our consciousness and our perception of reality.

But how and when did we begin to speak? What were the anatomic, genetic, mental, and social conditions that needed to be in place before we could utter our first word?

Long before Jean-Baptiste Lamarck, Charles Darwin, and Alfred Russel Wallace developed their theories of evolution, academics had spent centuries puzzling over the origins of language. In 1796, Johann Gottfried Herder, one of the most influential Enlightenment thinkers, published an essay in which he advanced the controversial argument that language was not a gift from God but a purely human invention—a point of view that angered many people at the time. After that, a heated debate about the origin of language arose—and continues to rage today.

From Bow-Wow to Ding-Dong

There were vast numbers of explanations of how language came to be, and they were all over the place. Most were pure speculation, quite a few were simply absurd, and a few garnered disparaging nicknames. The pooh-pooh theory speculated that language originated in automatic vocal responses to feelings such as pain, surprise, and fear: ouch! oh! or eek! According to the bow-wow theory, language came about as humans imitated the sounds in their environment onomatopoeically: the barks of dogs, the grunts of pigs, the rustling of leaves, the twittering of birds. This theory would explain, for example, why the word *hiss* sounds so much like the action it describes. The kling-klang theory maintained that the early ability to speak came from the involuntary sounds our ancestors made when they performed certain tasks, which then became associated with specific objects and individuals. According to this theory, the word *mama* comes from the smacking of the lips and the mmmm-sound infants make when nursing. The yo-he-ho theory started with the idea that the rhythmic grunts and chants used in cooperative, demanding physical tasks stand at the beginning of language. The la-la theory saw the origin of language in ritual dance, music, and incantations. According to the ding-dong theory, every object has a natural resonance, so every impression of the outside world elicits a characteristic sound in the head.[65] And this is just a small sampling of the theories…

Most of these theories about the origin of language started with the assumption that at some point, our early human ancestors began to make specific sounds that they associated with experiences, animals, plants, objects, and individuals in their environment. From there, they developed a rudimentary primitive language made up from short series of sounds spoken one after the other. These could have been grunts, howls, or hissing sounds that were always enunciated in the same way to stand for particular mental constructs. They would convey simple messages such as "Careful! Lions!" or "Snakes. Large. Dangerous!" The sounds would be accompanied by specific gestures and a particular facial expression. But even vervet monkeys—small, lively, savannah monkeys that live in the eastern and southern parts of Africa—can do that.[66]

From the perspective of current research, these theories are grossly oversimplified. On the one hand, they say nothing about the biology of speech and how it developed. On the other, they do nothing to explain how the basics of grammar, those rules that underlie the structure of every language, came about. Grammar is what allows us to connect individual words in a way that makes sense, to exchange complex content, and to point to things that happened in the past, are happening now, or will happen in the future. In his book *The Language Instinct*, the Canadian psycholinguist Steven Pinker uses a series of informative examples to show how critical grammar is for meaningful communication. It makes a difference, for example, whether there are animals around that humans can eat or animals around that might eat humans, or if you have found fruits that are ripe, that were ripe, or that will be ripe [67]

In Praise of Language

Grammatical structure goes far beyond the simple spoken forms of communication used by animals, such as alarm or location calls or information about places rich in food. Many animals can do that, from crickets to songbirds to monkeys to whales. Researchers have even taught a few highly gifted chimpanzees up to 250 different words using sign language or pictograms on a computer keyboard. The chimpanzees could string three to five words together to form a rudimentary sentence, but that was as far as they could go.

The problem with researching the genesis of language is that it must be deduced in a roundabout way using circumstantial evidence, because language use cannot clearly be tied to specific anatomical features found in fossils. The form and placement of the larynx could be a useful clue, but it is made of tissue that is not preserved in fossils. The only clues to the development of language that are sometimes found in fossils are the shape of the skull, the hyoid (tongue bone), the mouth, and openings in the skull for the nerves associated with the muscles used for speech. Hard facts like these, however, are a rare commodity in the search for traces of language.

The most diagnostically conclusive evidence is the increase in brain volume over the course of human evolution, and this can be reliably read from the shape of the skull. Most experts agree a sufficiently large brain is a necessary precondition of language acquisition. A large enough brain provides both the learning capacity necessary for mastering the multi-layered elements of language and the executive capacity necessary for coordinating the motor functions of the complicated apparatus of speech. To produce clearly articulated sounds that are easily distinguishable from each other, you need a finely regulated interaction of the diaphragm, tongue, teeth, gums, nose, larynx, vocal cords, lips, and over one hundred muscles.

By the end of its era, *Homo erectus* possessed a brain volume almost as large as ours. Its use of fire and the stone axes it fashioned so carefully show it was intelligent and capable of learning. But it is highly likely that these upright humans had significant deficits in the frontal lobes of the cerebral cortex. The frontal lobes are situated right behind our foreheads and are the area of the brain that organizes our thoughts. The frontal lobes are important for, among other things, language production, self-awareness, and the development of personality. In contrast to a modern human's skull, the skull of *Homo erectus* does not allow much space for this area of the brain. Almost all the increase in brain volume in modern humans took place in this area. It is unlikely, therefore, that

Homo erectus excelled in either smart remarks or in-depth conversation. Some experts, however, believe this species could communicate well using hand gestures and facial expressions.[68]

The Search for the Language Gene

Academic opinions today about when spoken language developed for the first time vary widely. Some experts talk of 100,000 years before today, others talk of 1 million years. Scientists such as the former curator at the American Museum of Natural History Ian Tattersall and the linguist Noam Chomsky believe symbolic thinking is a prerequisite for any language with a meaningful, logical sentence structure.[69] Symbolic thinking is the ability to translate things and experiences into abstract symbols such as gestures, sounds, images, or objects. Finds that can be construed as early forms of art are evidence of symbolic thinking.

The oldest find of this kind to date is a piece of ochre from the Blombos Cave in South Africa on which humans inscribed a netlike pattern eighty thousand years ago. According to Tattersall, the origin of language could not lie further back in time than 100,000 or maybe 200,000 years. Results from recent paleogenetic studies, however, point much further back in the past. In 2010, an early version of the Neanderthal genome was published.[70] Since then, there has been a series of informative comparative analyses between the Neanderthal genome and the genome of modern humans. The researchers found FOXP2,[71] commonly known as the "language gene," particularly interesting with regard to the ability to speak.

FOXP2 was discovered back in the 1990s when it turned out that the severe hereditary speech disorders in a London family could be traced back to a defective gene. Astoundingly, this defect was handed down over three generations. Half the family members had difficulties putting their thoughts into words, as well as problems formulating and understanding speech. FOXP2 is also found in numerous mammals in slightly modified forms. One modification to the gene in laboratory mice led to severe disruptions in movement and behavior,[72] and the animals also could not understand each other when they tried to communicate using ultrasonic sounds.[73] Since then, we have learned that FOXP2 plays a significant role in the human genome when language is inherited—including the ability to use grammar—and that two functional copies of the language gene are needed to pass down normal speech.

Chimpanzees, the closest relative of *Homo sapiens*, possess a FOXP2 gene that differs from the human gene in two distinct places. In orangutans, it differs in three places. Modifications in the gene, therefore, could dictate different trajectories for language proficiency over the course of human evolution. It seems, then, that it would be a good idea to make a comparison with the Neanderthal genome. Surprisingly, it turns out that FOXP2 in Neanderthals and in humans is almost exactly the same.

Scientists at the Max Planck Institute for Psycholinguistics at Nijmegen in the Netherlands have spent years investigating all available genetic, anatomical, and archeological indicators that might give information about when our ancestors started to speak.[74] After evaluating the extensive data, they concluded that in all likelihood Neanderthals could speak and that human language arose at least as far back as the last common ancestor of modern humans and Neanderthals. This last common ancestor is probably *Homo heidelbergensis* (Heidelberg Man), whose oldest finds are at least 600,000 years old, or perhaps a more advanced version of *Homo erectus*. This idea is supported by the analysis of the language gene, and also by, among other things, the thickening of the thoracic spine at this time for better enervation of the muscles that control breathing, a widening of the canal for the nerve that controls the tongue, as well as an increase in the area of the cerebral cortex that is responsible for motor control and where, most importantly, the contents of perception are processed. In addition to this, the shape of the hyoid or tongue bone, which is critical to speech production, is very similar in Neanderthals and modern humans.

With "the accumulation of [both genetic and cultural] small changes," the scientists wrote, language slowly developed over a very long period of time. They concluded that "Neandertals, Denisovans and contemporary modern humans shared a similar capacity for modern language, speech and culture."[75] From their perspective, early steps in preparation for language could have been taken over a million years ago. And so it is not out of the question that *Homo floresiensis*, probably the first seafarer in the history of human evolution, had a basic command of language. Without language, it would never have managed to prepare for the journey.

The Value of Cooperation

We know how important language is for us because there is a special genetic program dedicated to learning it. The critical stage for language acquisition lasts from four months until about the age of ten. During this phase, we soak up language and play around with it until we master it. After that, language acquisition becomes much more difficult and we have to study long and hard to learn vocabulary and grammar. What is it about language that is so incredibly valuable for us that we have developed our own fast-track learning program?

It is commonly thought that of all the things humans did, speech was most useful to us when it came to exchanging information about hunting and gathering food, and this is why the ability to speak was given priority in the evolutionary selection process. But, of course, the exchange of information is just one of the ways we use language. There are many others.

The British anthropologist and evolutionary psychologist Robin Dunbar at the University of Oxford was researching grooming in primates when he came across a correlation between the size of the neocortex and the size of social groups. Chimpanzees live in groups of about 50 individuals, whereas humans tend to be part of social circles of about 150 people. "Dunbar's number," as it is known, corresponds to the size of villages in early tribal societies, but also to a large extent to the number of important people in our modern social networks today. Above a certain group size, Dunbar says, mutual grooming no longer works as a form of nonverbal communication that holds groups together. You can comb through the fur of only one other individual at a time, but you can talk to many of them all at once.

An increasingly complex group dynamic calls for better skills in interpreting the thoughts and emotions of others. Dunbar thinks that in the transition from small groups of apes to early human communities, plenty of gossip was the social glue that made this type of human coexistence possible. Gossip facilitates an exchange of emotions within a group. It is mostly about what behavior is acceptable and what transgresses the accepted norms, who is honest and who steals from others, who cannot abide whom, or who is sleeping with whom.[76] Even in the modern world, more than 60 percent of what we talk about on a daily basis concerns interpersonal relationships. And these kinds of conversations are the ones we excel in. It is much easier for us to talk about the foibles of our neighbors than to give clear instructions on how to tie a shoelace or install a faucet. When it comes to small talk, the tone of voice and the rhythm of the conversation are often more important than content, because what is really important is making contact on an emotional level.

More than any other creature, we are interested in what is going on inside the heads of others. Thanks to our well-developed sense of empathy, we can figure this out often enough. We are masters of detecting plans, motivation, and intent. And we are eager to share experiences, interests, and rules with others.

In numerous studies comparing young children and chimpanzees, Michael Tomasello—the anthropologist and behavioral researcher mentioned in Chapter 18—observed that children were more advanced than the chimpanzees in two ways: they were much better at putting themselves in the shoes of their playmates and developing a sense of self, and they spontaneously suggested mutual cooperation. "Give, and it shall be given unto you" is not something you have to teach children; they are born thinking that way. Unlike our nearest relatives, most members of our species are capable of seeing the world from the perspective of "we," even early on

in our personal development.[77] Tomasello calls this skill "cultural learning." The development of spoken languages honed
this skill and helped us become increasingly social creatures.
Tomasello sees the roots of language development as part of
our immense investment in socialization, and sees speaking
as an evolutionary accomplishment that makes it much easier for us to be social creatures.

Modern humans can be empathetic and selfless with
close family, circles of friends, and interest groups such as
sports teams or neighborhood swap meets, but they can
also be unwelcoming and hostile to groups that do not share
their interests, appearance, or language. Within our trusted
in-group, we court sympathy and fight for recognition and
good standing. We cut other groups out. We have inherited
these traits from the dim and distant past, when competition
for food, mates, and other resources was more important to
our survival than it is today. This is the dark side of social
skills that need to be more fully developed. However, with
our ability for self-reflection and the opportunities we have
to speak to each other, in principle, we have at our disposal
the means to overcome such divisions.

The interplay of language and consciousness ushered in
another aspect of becoming human, one that finally raised
Homo sapiens above the animal realm: the opportunity to
develop human civilization. With our language skills, we
could now share substantial amounts of information within
a large group. Unlike biological evolution, where changes are
handed down through individuals over generations, our cultural evolution developed at an enormously accelerated pace,
because knowledge could now spread rapidly within a group.

Of course, animals, too, can hand down information about

experiences they have had and about what has been done in the past. Birds sing in local dialects. Whales develop specialized techniques for catching fish, which they hand down to the next generation. In some groups of chimpanzees, individuals share their experiences about the beneficial effects of certain healing plants. However, all of this happens in the severely circumscribed context of demonstrating a behavior that others then imitate. Language in conjunction with cave paintings, music, and rituals elevated the exchange of information to a whole new level. Now experiences, knowledge, and world views were preserved in stories and songs. When written language evolved about five thousand to six thousand years ago in the Fertile Crescent between the Tigris and Euphrates Rivers and word of this technique for preserving culture got around, the amount of knowledge that could be handed down exploded, because it was now independent of the storage capacity of the human brain.

At first, all the people in Mesopotamia wanted to write down were harvest records and royal edicts. Then, as writing spread and the techniques for capturing characters were refined, the floodgates opened and educated people began exploiting this new medium in a wide variety of ways. Writing was increasingly no longer about record-keeping and laws, but about poetry, stories, epic adventures, plays, medical works, construction manuals, encyclopedias of knowledge, and holy texts. There was another huge leap forward when the printing press was invented, and then another when digital data became commonly accessible. Without language and writing, there would have been no technological, social, and cultural innovations, no rockets to the moon, no social safety networks, no Bach oratories—and no atomic bomb.

WHAT BEGAN WITH *Danuvius* and *Graecopithecus* millions of years ago with the gradual development of upright gait and the freeing of hands, led much later—thanks to better nutrition and the development of language—to the complex social structures of us, modern humans. But it is only in hindsight that this process looks like a straight path toward a defined goal. In reality, it was an evolutionary network involving many species. Of all the species of humans that once settled this planet, only one remains: *Homo sapiens*. Why was this the only one that survived?

Part 6

THE
LONE
SURVIVOR

- 23 -

A CONFUSING
COMPLEXITY

The Problem With the Family Tree

IDEAS IN PALEOANTHROPOLOGY that were considered irrefutable for a long time have recently been thrown into disarray and must now be recalibrated. Numerous new finds present puzzles that are difficult to fit into the overall picture. The boundaries between the individual human species are fluid and not always easy to define. Research cannot yet show a clear, conclusive line of descent from early hominins to us. Today, the path to becoming human since we split from chimpanzees looks less like a family tree and more like a braided river delta, where some channels take their own course only to reconnect later, while others become rivulets before, at some point, disappearing completely.

Science's position with regard to the genus *Homo*, that is to say our closest relatives, has become particularly difficult to follow of late. Today, there is not even consensus on how many species of human there were. Some ideas about which criteria should be used to differentiate the individual species diverge widely from others. That is partly because the transitional steps between the often-incomplete finds of different species are unclear. And partly because there are pronounced anatomical differences within individual species. It would be like scientists in the distant future finding the bones of the 7-foot-tall (2.13-meter) basketball giant Dirk Nowitzki and those of the 5-foot-7-inch-tall (1.69-meter) soccer player Lionel Messi next to each other and wondering if they were looking at anatomical variants of a single species or at two separate species.

Some scientists take a radical view and recognize only *Homo erectus* and *Homo sapiens* as confirmed human species. The majority of paleoanthropologists today, however, are of the opinion that there are twelve species of the genus *Homo* that can be more or less clearly separated out.

Homo habilis
H. rudolfensis
H. georgicus
H. ergaster
H. erectus
H. antecessor
H. heidelbergensis
H. naledi
H. floresiensis
H. luzonensis
H. neanderthalensis
H. sapiens

This is confusing, and not only for lay people. Some fossil finds are categorized under two or three different names. A partial skeleton of a human that was less than 300,000 years old, found at Broken Hill in Zambia, has been classified under four species names: *Homo rhodensiensis, H. arcaicus, H. heidelbergensis,* and *H. sapiens.* I would like to attempt to bring some order to this multiplicity of names.

The largest gaps are found in the oldest members of the human genus, which is why, in this book, I have often used the term "early *Homo*" to simplify matters. Their finds date back to 2.5 million to 1.44 million years. They come from Ethiopia, Kenya, and Malawi (*Homo rudolfensis*), Kenya and Tanzania (*Homo habilis*), and possibly also China (*Homo wushanensis*). These were all relatively small humans, up to 5 feet (1.5 meters) tall and weighing up to 110 pounds (50 kilograms),[1] with a brain volume of 580 to 820 cubic centimeters. The smallest species is *Homo habilis,* which stood about 3 feet (1 meter) tall and weighed no more than 77 pounds (35 kilograms). It is possible that this species is still an early hominin, because many of its anatomical features are very similar to those of the genus *Australopithecus.* These features are also evidence of a well-developed ability to climb trees.

Current thinking holds that the oldest early humans (early *Homo*) were the first to make tools. Finds of stone tools dating back at least 2.6 million years have been found not only in eastern and northern Africa, but also more recently in Israel, Russia, India, and China. Because there is evidence that members of the genus *Australopithecus* already had deft hands, it is possible that the creators of some of these tools were early hominins.

Homo luzonensis from the Philippines and *Homo floresiensis* from Indonesia were anatomically closely related to *Australopithecus* and early *Homo.*

266 | ANCIENT BONES

Puzzling Early Humans From Georgia

Homo georgicus is a particularly intensively studied species of early human. It is between 1.85 million and 1.77 million years old and was discovered at a site in Dmanisi, Georgia. Unusually well-preserved partial skeletons with skulls from five individuals give us the most complete picture we have of early humans from the beginnings of the ice age.

The most peculiar thing about this early species of human is its combination of primitive and advanced features. Its relatively small brain is primitive. At a volume of 550 cubic centimeters to 750 cubic centimeters, it is about the same volume as that of an early *Homo*. And standing about 5 feet (a little over 1.5 meters), it is also relatively small. The structure of its upper arm is also primitive. When *Homo georgicus* let its arms hang down by its sides, its palms faced forward like those of great apes, early hominins, and *Homo floresiensis*, instead of toward its thighs like modern humans. The anatomy of its lower extremities, however, was much more highly developed, and was already similar to that of *Homo sapiens*. Its legs were long and straight and its foot had an elastic arch, making *Homo georgicus* the oldest known "running" human.

None of that fits with the accepted Out of Africa I theory. After all, according to that theory, the first early humans outside Africa would be little more than a million years old, stand over 6 feet (1.7 meters) tall, and have a brain volume of at least 1,000 cubic centimeters.[2] These assumptions were completely contradicted by the finds from Dmanisi.[3]

There is one more aspect of these fossils that is noteworthy. One of them was an old man who had lost all but one of his teeth, and the degeneration of his jaw suggested he had survived without them for years.[4] It is impossible to imagine how this old man would have managed to stay alive over 1.8 million years ago without a social support network. Ethnological studies have shown that Indigenous people in similar situations are fed food chewed by other members of their group, an act that strengthens the resilience of seniors and demonstrates that the helpers possess a high level of social competence and a sense of empathy—a behavior that is truly human.

The skull of the early human *Homo georgicus* from Dmanisi, Georgia

For some researchers, *Homo georgicus*'s anatomical and social features, combined with its advanced geological age, are good reasons to put the Georgian early human forward as a leading candidate for direct ancestor to the archetypal early human from Asia, *Homo erectus*. Up until now, *Homo erectus* has been the model for the longest-existing human species, as it lived on Earth for at least 1.5 million years, by far the longest period of time of any human species documented so far.[5]

Ice Ages, Savannahstan, and the Old World

As mentioned, we do not have enough fossils to adequately explain the transition from early hominins of the genus *Australopithecus* to early humans of the genus *Homo*, so there is a lot of room for speculation. One thing, however, is clear: the transition took place between 3 million and 2 million years ago. Because of the dearth of fossils, there is debate about what spurred this development and where it took place. In all the currently accepted theories, the appearance of early *Homo* centers on Africa, because Africa is the only place where australopithecines have so far been found. Also taking up a lot of room in this debate is the importance of climate change as a driver of evolution. The transition 2.6 million years ago from the overwhelmingly warm Pliocene to the fluctuating cycles of the ice age was doubtless one of the most dramatic changes in climate in Earth's recent geological history. This change marked the hour of the birth of our genus and led to the evolution of the archetypal early human, *Homo erectus* (Upright Man).

The characteristic features of the early ice age were glacial periods about every forty thousand years that covered the high mountain ranges and the arctic regions with ice.[6] When glacial periods locked up water in ice sheets at the poles and in glaciers, sea levels dropped and there was drought in temperate zones. These climate changes were caused by shifting continental plates, which altered the course of ocean currents and increased volcanic activity. These changes are driven by periodic fluctuations in solar radiation as the Earth makes its elliptical orbit, and the changes are also influenced by the tilt and wobble of the Earth's axis.

During the glacial periods, sea levels dropped drastically, by up to 400 feet (120 meters). Winters were cold, the climate was dry, and grasslands expanded. In the warmer interglacial periods, when the temperatures in some places were much higher than they are today, it was relatively humid, and sea levels were much higher once again, in places up to 165 feet (50 meters)[7] above present-day levels.

As the cold periods intensified, deserts expanded. Enormous quantities of loess, a sedimentary deposit made from fossil dust, began to accumulate starting 2.6 million years ago in northern China and 2.4 million years ago on the shores of the Caspian Sea.[8] Deserts such as the Gobi and the Karakum expanded and retreated in rhythm with the global climate. They joined up with deserts in North Africa and Arabia to create the Old World desert belt that still stretches today from the Atlantic coast of Mauritania to Mongolia.

Some scientists believe that these inhospitable deserts restricted the early evolution of the genus *Homo* exclusively to East and South Africa. Yet the cyclically changing landscapes on the edges of the deserts could have accelerated early human evolution. There, forest-rich habitats morphed into bush savannahs and grasslands relatively quickly. These changes were particularly pronounced in the Mediterranean region and in a region that stretched from eastern Europe to Central Asia. In habitats like these, where change was constant, the only early humans that could survive were the ones who knew how to adapt—whether that be through inventions such as tools or discoveries such as the usefulness of fire. Under these unstable living conditions, selection would have kicked into high gear. Intelligence, creativity, and flexibility would become ever-more important to ensure survival.

These adaptations were crucial for early humans since a changing climate was not the only challenge they were facing. They also had to cope with the dramatic changes in the animal and plant world that accompanied it, particularly in Asia and Europe.

At the beginning of the ice age, horses appeared for the first time in Eurasia. Zebras and donkeys are among their descendants today. The grasslands and savannahs were just the kinds of habitats they were made for. Horses in the genus *Equus* originally appeared in North America. They made their way to Eurasia via a land bridge that formed across the Bering Strait from North America to Siberia when glaciation caused sea levels to drop. It did not take them long to spread all over the Eurasian land mass, and 2.3 million years ago they arrived in Africa. Other animals that came over from North America included the forerunners of wolves, jackals, and coyotes, which appeared in the Old World 2.1 million years ago. Mammoths, *Mammuthus*, a genus that originally appeared in Africa, traveled in the opposite direction. Shortly before the beginning of the ice age, they expanded their range to the whole of Eurasia and later made it to North America.[9]

The mammal fauna during the ice age, especially in the Mediterranean region, was a mix of species from both Eurasia and Africa. Goats, bears, and raccoon dogs were just as at home in North Africa as giraffes and mammoths were, while in southern Europe, there were gelada baboons and hippopotamuses.[10]

Given the extensive integration of northern and southern ecosystems, the archeologists Robin Dennell and Wil Roebroeks advised against viewing Africa, Europe, and Asia as discrete, independent systems.[11] Instead, they argued, the boundaries between these continents were in flux. For example, there were savannahs and plains where grasses and herbaceous perennial plants were the dominant form of vegetation on both sides of the Old World desert belt. Dennell and Roebroeks therefore suggested a new term to connect them: Savannahstan. This description includes the whole grassland ecosystem north and south of the desert belt, stretching all the way from North Africa to East Asia and shaped by global climate. Perhaps it was not Africa but Savannahstan—which includes large parts of Eurasia—that was the cradle of humanity? The focus on one particular continent, given all the ecological, climatic, and evolutionary connections between Eurasia and Africa, is simply too narrow. And the enigmatic Denisovans are proof of this.

– 24 –

A PUZZLING
PHENOMENON

Humans From Denisova Cave

I N THE SUMMER of 2008, Russian scientists scored an
unexpected coup that stunned more than just the academic world. Two researchers from the Russian Academy
of Sciences, Michael Shunkow, a paleontologist, and Ana-
toly Derevianko, an archeologist, were conducting a dig
in a remote valley in the Altai Mountains in southwestern
Siberia.

About 90 feet (28 meters) above the Anui River, in the
middle of a picturesque landscape, there is a cave that until
then was known to only a few experts: Denisova Cave. Leg-
end has it that the name can be traced back to a settler called
Denis who lived there in the eighteenth century. Scientists
had been digging here since the 1980s, and over the years,
researchers had dug deep down into the clay deposits on the
cave floor. They kept coming across tools and pieces of jew-
elry. This suggested that the cave had been visited regularly

by humans for a long time, first Neanderthals and then modern humans.

Along with the handcrafted objects, over time the scientists excavating the site had also unearthed thousands of bone fragments. The vast majority were the remains of animals: ice-age hyenas and cave lions that had been killed, dragged into the cave, and eaten there. Many of the bones were so fragmented that it was not possible to tell exactly what animals they came from.

Because the individual pieces of the ice-age puzzle in general are usually very small, during the 2008 digging season, the researchers were digging with utmost care, just as they had done in all other seasons. The light conditions in the cave are not good. Important fragments could easily be overlooked. And so they removed the sediment carefully, packed it into wooden crates, and used a rope and pulley system to lower the crates down to the river, where the sediment was washed in fine sieves to release the tiniest of fragments of bone.

That summer, Shunkow and Derevianko were rewarded for their painstaking work. They came across a tiny fossil that looked familiar and quickly ascertained that it was the bone at the end of a human little finger. However, the anatomical information contained in a single finger bone is not sufficient to prove it belonged to a human. For the researchers, there were only two options: Neanderthal or modern human. Both species lived in this area in the late ice age. There was proof of this not only from finds in Denisova Cave but also from fossils and artifacts from other digs in the Altai Mountains. The Russian researchers therefore decided to do a genetic analysis of the tiny piece of bone, and they sent it to the Max Planck Institute for Evolutionary Anthropology in Leipzig.

The Complete Genome of an Early Human

At the time, the paleogeneticist Svante Pääbo and his team were working at the institute trying to decode the Neanderthal genome. Pääbo had already developed a method to extract genetic material from bones as ancient as this one. He would extract the material, clone it, and put it back together again in a process known as gene sequencing. The first thing the experts did was bore about 30 milligrams of bone from the tiny Siberian fossil. Their goal: to extract a special form of genetic material known as mitochondrial DNA. This part of the genetic material is found not in the nucleus of the cell but in tiny organelles in the cell, the mitochondria, that are responsible for providing the cell with energy. They are often also called the cell's powerhouses. It is easier to decode mitochondrial DNA than to reconstruct all the genetic material in a cell's nucleus, and mitochondrial DNA is sufficient to classify the fossil as belonging to a human.

A critical factor affecting the success of sequencing is how well preserved the fossil is. The colder the conditions at the site over thousands of years, the better. Even today, temperatures in Denisova Cave never rise above 44.5 degrees Fahrenheit (7 degrees Centigrade). The conditions needed to extract genetic information from the finger bone seemed to have been met, and no one was expecting any big surprises from the analysis.

In January 2010, Pääbo received an unexpected phone call from his colleague Johannes Krause, who had analyzed the bone. Krause had sensational news: the genetic fingerprint from the fossil belonged to neither a Neanderthal nor a modern human. Pääbo's first thought was that his colleague had made a mistake. But it quickly became clear that the result was correct. The finger bone came from an unknown form of human. The scientists had discovered a new extinct human relative. Immediately, the researchers decided to decode all the genetic material in the nucleus of the cell. They took another tiny core of bone and got to work. Success.

Finally, in 2012, the research team published the complete genome of this new member of the human family tree and called the group Denisovans. At first, they tried to call the new species *Homo altaiensis*. But because of difficulties in pinning down an exact biological boundary—a problem that exists with Neanderthals as well—they rejected this idea. The genetic analysis showed that Denisovans were more closely related to Neanderthals than to modern humans, but that they had had a long independent evolutionary history of their own. The last common ancestor of Neanderthals and Denisovans lived about 450,000 years ago, and the last common ancestor of Denisovans, Neanderthals, and modern humans lived as far back as 800,000 to 600,000 years ago.[12]

The researchers, however, wanted to find out more about the relationships between the three groups. After years of analysis, they had just decoded the Neanderthal genome and determined that about 3 percent of the DNA in people living today is Neanderthal. African individuals carry a stronger signal for Neanderthal ancestry than previously thought, but distinctly less than people living in Asia or Europe.[13]

Could it be that the Denisovans also left genetic traces in modern humans? The researchers compared the Denisovan genome with the genetic material of people living today in many different parts of the world and made another fascinating discovery. The Indigenous people in Papua New

Guinea, Australia, the Solomon Islands, and members of a few tribes in the Philippines all carried up to 5 percent Denisovan DNA.[14] But these regions are thousands of miles from Siberia.

Here is the most likely explanation: At that time, the human population of Eurasia split into two groups. The Altai Mountains likely formed the western boundary of the area settled by the Denisovans, and the Neanderthals likely never got any farther east than this. Whereas the Neanderthals mostly lived in Europe and the Near East, the Denisovans likely settled the rest of Asia.

Ménage à Trois

Clearly the Denisovans even managed to get to islands. This was an amazing achievement, similar to that of the Hobbits and the early humans on the island of Luzon.[15] Almost no one would have thought it possible so early in the story of human evolution. If you assume that modern humans and Denisovans first interbred in Australia and New Guinea— and some researchers think this is entirely possible—then the Denisovans would have had to cross the Wallace Line[16] (see the diagram on page 223), as *Homo floresiensis* had done before them.

The Wallace Line is a biogeographic boundary in Southeast Asia. Even when sea levels were at their lowest during the ice age, there were still ocean straits along this line, which meant it was almost impossible for an exchange of animals to happen here. The line runs from today's Indonesian islands of Bali and Lombok, continues farther north between Borneo and Sulawesi, and reaches as far north as the southern

Philippines. It divides the very different animal worlds of western South Asia on the one side and Southeast Asia, Australia, and Oceania on the other.

Ocean straits like this were a big barrier to the early migration of humans. But a recent study by an international team of researchers suggests that the Denisovans overcame them.[17] They came to the conclusion that people who live in Papua New Guinea today carry genetic material from two different evolutionary lines of Denisovans that split 350,000 years ago. Their results also show that a population of Denisovans might have lived there as recently as 30,000 years ago.

What is certain is that modern humans and Denisovans met in many different regions and had children together. In 2018, another find from Denisova Cave made the headlines: a story of Stone Age interbreeding between two different kinds of humans added a new chapter to the history of humankind. The find was entirely due to the perseverance of a few experts from the University of Oxford.[18] Over the years, they had developed a technique that allowed them to identify fragments of bone with great precision, even if they were tiny and badly disintegrated. Experts used mass spectrometers to get a sort of molecular fingerprint stored in the collagen in the bone that differs from animal to animal and between humans and animals. This technique has the added advantage that you can use it to evaluate many samples in a short period of time.

To test their method, the researchers received a sack of unidentified bones from Denisova Cave from their Russian colleagues, and they embarked on trying to find human remains. After the first test run with 700 samples that did not show any results, they did a second run with 1,500 more samples and finally found a match: a human bone about 0.75 inches (2 centimeters) long. But it was not clear which kind of human it belonged to. And so the paleogenetic research- ers from Leipzig were once again called on for help. And for a second time, their investigation led to a completely unexpected answer: Denny. She was the daughter of a Nean- derthal mother and a Denisovan father, the only fossil of a direct hybrid of two different types of humans ever found.[19]

Denny was the product of a liaison of the two types of humans 100,000 to 90,000 years ago, and she was at least thirteen years old when she died. Her genetic material shows not only that Denisovans and Neanderthals could clearly have offspring together but also that it was not unusual for them to do so: the researchers discovered that one of the Denisovan father's ancestors was a Neanderthal. The investi- gation also discovered that Denny's Neanderthal mother was more closely related to west European Neanderthals than to those found in the Altai Mountains. That means Neander- thals must have covered great distances when they traveled. And there we have it. A fascinating story of Neanderthals and Denisovans living in close proximity, and of their interbreed- ing and migration, all written in a tiny, nondescript piece of bone.

Now that the complete genome of the Denisovans has been decoded and published, other researchers around the world are using the data for comparative studies with today's humans. One outcome of this was a study published in 2014 that suggested that a variant of the EPAS1 gene, also known as the super-athlete gene, could be an archaic inheritance from Denisovans.[20] Modern Tibetans are the people most likely to carry this gene. It allows their bodies to process oxygen efficiently at high altitudes, which means they can undertake physically demanding tasks even in thin mountain air and are well equipped to live life in areas above 13,000 feet (4,000 meters). The most famous example of people who carry this gene are Sherpas. Mountain climbers hire them because they are extremely fit and can carry heavy loads as porters on treks through the Himalayan peaks. People who do not possess this gene variant produce more hemoglobin and red blood cells at altitude to compensate for the lack of oxygen. That makes their blood thicker, which greatly increases their risk of forming blood clots. In the worst-case scenario, these clots could lead to strokes or heart attacks.

In a 2016 study, a separate group of researchers concluded that Greenland Inuit might also carry gene variants that can be traced back to Denisovans.[21] In this case, the genes handed down allowed the human body to store a particular kind of fat that it burns to keep warm. This was probably very advantageous to Denisovans living in the ice-age climate. After the ice age, these genes also helped modern humans living in cold regions to survive and were therefore selected for and handed down mostly in those regions.

The finds from Denisova Cave added a fascinating new chapter to the story of human evolution and will probably lead to more surprises in the future. As yet, however, they have given us no firm clues about what Denisovans looked like. Did they look more like the Neanderthals with their muscular compact bodies, thick bony ridges over their eyes, and flat, sloping foreheads? Or were they anatomically more similar to us, *Homo sapiens*? Since the initial Denisova Cave finds, three more teeth found in the cave and a jawbone found in Tibet have also been identified as Denisovan. However, we will not be able to do an anatomical reconstruction until more bones or, with any luck, a skull are found. Until that happens, the Denisovans will remain shadowy figures that once peopled a large part of Eurasia but left few traces behind.

– 25 –

AND THEN THERE WAS ONE

The Rational Human

———

THE MORE THAN 7 billion people who inhabit the planet today all belong to one species of human: *Homo sapiens*, which can be translated as Rational Human. When this species first stepped onto the evolutionary stage over 300,000 years ago, it was by no means the only species of human on Earth. As far as we know, it shared Eurasia and Africa with seven other species of human.

Eurasia was settled by the Denisovans, the Neanderthals, *Homo heidelbergensis*, and *Homo erectus*. *Homo floresiensis* and *Homo luzonensis* lived on islands in Southeast Asia, and *Homo naledi* lived in South Africa. From an evolutionary perspective, however, these species were not to coexist for long. Something happened over the course of about fourteen thousand human generations to change things, but what? Why did the diversity of human species shrink so rapidly and so drastically? Why, of all these species, was ours the one to prevail? Were the fellow humans with whom we shared the Earth overtaken by the same fate as the large mammals, the so-called megafauna, of the ice age? Did *Homo sapiens* drive them to extinction?

The species of megafauna varied from continent to continent, but these carnivores and herbivores were all impressively large. In Eurasia, for example, they included mammoths, cave lions, wooly rhinoceroses, and giant deer standing more than 6 feet (2 meters) at the shoulder. In the Americas, there were giant sloths, saber-toothed cats, elephant-like mastodons, and steppe mammoths. In Australia, there were marsupial lions, giant wombats, 10-foot-tall (3-meter) kangaroos, and thunderbirds—chunky-looking dromornithids that were up to 10 feet (3 meters) tall and weighed up to half a ton. All of them died out during the past forty thousand years.

These mighty animals were not doomed by climate change alone. Modern humans also played their part. These megafauna were slow to reproduce and could not survive the relentless pressure applied by modern humans, who hunted them over a long period of time using increasingly sophisticated methods.[22] The only places a few particularly large mammals such as elephants and rhinoceroses survived were Africa and, to a slightly lesser extent, southern Asia. Today, however, they are severely threatened by habitat loss and hunting.

Way back in the 1960s, in his overkill hypothesis, the American paleontologist Paul Schultz Martin already suspected that modern humans were responsible for the extinction of the megafauna.[23] The disappearance of these animals coincided with the appearance and spread of *Homo sapiens* but not with other species of human. Is it ingrained in modern humans to efficiently exploit food sources until they are gone, and to destroy anything that could threaten them or compete with them, groups like Neanderthals and Denisovans, perhaps? I do not think so. The reasons to take a completely different view of what happened lie hidden in our genetic code.

The Early Human Within Us

Genetic analysis of the human population alive today does not give us a clear picture of the history of *Homo sapiens*' lineage. There are two main reasons for this. First, there were human populations that left either no or very little genetic material in living humans. It could be that these populations either died out without leaving any descendants or their particular genetic material was not selected for and they therefore died out. Second, the people who live in a particular region today are rarely genetically related to those who settled the region thousands of generations ago. For example, the distant descendants of the first crop farmers in Europe are no longer on the mainland, but can be found in Sardinia. The descendants of the Paleolithic hunters and gatherers of eastern Siberia now live in the Americas. Researchers have found the genes of Denisovans in populations in New Guinea and Australia. The key to understanding more recent human history, therefore, lies in paleogenetic data extracted from the bones and teeth of extinct populations.[24]

Paleogenetic research shows that about 2 percent to 8 percent of the genome for our species, *Homo sapiens*, comes from other species in the genus *Homo*. Genes from early humans can be found in all modern humans, whether they live in

Africa, Eurasia, Australia, or the Americas. These genes do not always show up in the same portion of the genetic code, and they often come from different species of humans.

Geneticists have found gene sequences from at least five different species of humans in humans living today, each with a characteristic geographic distribution.[25] Europeans carry about 2 percent Neanderthal genes within them.[26] Asians and the Indigenous populations of the Americas carry genes from both Neanderthals and Denisovans. Up to 8 percent of the genetic material carried by Melanesians from the Philippines, New Guinea, and Australia comes from a total of four human species: Neanderthals, Denisovans, and two other species that scientists have not yet been able to identify. Finally, Africans carry evidence of one or two other currently unidentified human species.[27]

These genetic contributions can be traced back to interbreeding between the different species, despite distinct differences in appearance and behavior. But that seems not to have bothered the different species of humans very much. And they were still genetically so similar that relationships resulted in healthy offspring that could themselves reproduce. Such pairings happened not only between a number of different species, but also repeatedly within the same species at different times and in different geographic regions.[28]

This genetic information from early humans occupies different portions of our genome and guides a variety of processes. Neanderthal genes have been linked to brain development[29] and neuronal function.[30] Denisovan genes have been found mostly in portions of the genome that control the growth of bones and tissue.[31] The knowledge we have gleaned from paleogenetic studies over the past twenty years leaves no doubt that different species of humans interbred. Moreover, these liaisons were not the exception but the rule, and they were essential for the rise of the often-variable and adaptable species that today we call *Homo sapiens*.[32]

Not Guilty but Still to Blame

As all living humans carry genes from early humans, albeit different genes in different people, this means that our genome has retained many genes belonging to earlier humans who trod our evolutionary path, and today these genes are widely distributed within the human population. At the moment, experts believe that 30 percent of the complete Neanderthal genome is retained in the non-African population and maybe as much as 90 percent of the Denisovan genetic material.[33] Can we therefore say that these earlier fellow humans went extinct? Or would it be more accurate to say that they became part of us and live on in us today?

Every one of us carries a small piece of them within us. Their genes are an essential component of the genetic material of humans today. *Homo sapiens*, therefore, did not drive the other species of humans that once lived with us on this planet to extinction; instead we simply merged with them. Modern genetics absolves us of the role of ruthless murderer who swept every non–*Homo sapiens* species of human out of the way. And yet, we cannot absolve *Homo sapiens* of driving many animal species to extinction. On the contrary, since our species appeared on Earth, a multitude of species have disappeared because of human actions. What began catastrophically has only continued to get worse. The Intergovernmental Science-Policy Platform on Biodiversity and

Ecosystem Services (IPBES) estimates that over 1 million species of animals and plants stand on the brink of extinction today as a result of human interference in nature. The largest extinction of species since the mass extinction at the end of the dinosaur era over 65 million years ago is looming on the horizon.

And here is one more thing to give us pause for thought as we look back over the path that led to our becoming human. The base pairs of the complete genomes of two Neanderthals, one from Croatia and one from southern Siberia, had differences of only 0.16 percent.[34] Although the two lived 3,500 miles (5,500 kilometers) apart, they were genetically almost identical and more closely related to each other than two people chosen at random from today's population of 7 billion people would be. That suggests a very small population density for Neanderthals, a group that settled a vast region. Paleogeneticists estimate that the effective population size— that is, the number of people of reproductive age needed to maintain population levels—of Neanderthals and Denisovans would be only about two thousand to three thousand individuals. The effective population size for *Homo sapiens* at this time was about five times larger.[35] Below this critical level, there was an increased risk of genetic defects if men and women who were too closely related had children together. Genetic defects would lead to falling fertility rates and rising death rates.

If the number of Neanderthals who lived in Europe was only a few thousand, there might have been only a few hundred of these hunters-gatherers in Europe in the harsh conditions of the last glacial period.[36] The small population size was what led to their demise. Even in favorable climate conditions during the warm interglacial periods, no more than half a million to a million humans lived in Eurasia, and they were not all of the same species.[37] These few men, women, and children divided up into small groups of twenty-five to thirty individuals and spread over a huge area from Britain to Kamchatka and Siberia to India. Many studies suggest that at the end of the last glacial period, roughly twelve thousand years ago, only about ten thousand humans were still living in Eurasia. It is likely that there were only about 1 million people living on this blue planet at that time. If just one little thing had gone wrong, the human experiment might have ended right there.

The situation did not change until, about ten thousand years ago, modern humans invented livestock husbandry and crop farming. After people began settling down in one place and farming, the human population increased to many millions for the first time—and humans began to have an increasing influence on their environment. These inventive, "rational" people cleared forests, domesticated animals, and raised crops—and hunted more and more species to extinction. Humans began to feel they were no longer part of their environment and created their own urban realm. Humanity reproduced until there were billions of them and their need for resources grew exponentially. This process has advanced so far in the last seventy years that scientists are now talking of a new epoch in Earth's history: the Anthropocene, or time of humans. Humans have raised themselves above the animal realm. And it seems the consequences will be dramatic.

The danger for humanity does not lie in climate change alone. Humanity is much more threatened by crisis situations that have developed in all spheres of the Earth's systems: the atmosphere, the oceans, the animal and plant world, and the soil. In my view, two developments are especially danger-ous: the destruction of natural habitats and the increasingly aggressive approach to what resources remain. If human-ity does not succeed in halting its population explosion and establishing an economy that is not based on growth, it is headed for catastrophe. It will not be a climate catastrophe, but first and foremost a social one. But that is another story, a story for another book.

EPILOGUE

H UMAN EVOLUTION IS currently one of the most dynamic fields of paleoanthropological research. Many discoveries are presented each year, and we have many new scientific techniques that allow us to reinvestigate finds made in the past. Paleoanthropology is also now happening in more countries and involves scientists from more disciplines than ever before. As a consequence, many finds have been made in unexpected and previously unexplored areas of Europe and Asia. It is therefore not surprising that long-held beliefs about why, how, where, and when the human lineage evolved are now being challenged.

This book focuses on some recent discoveries and what they mean in terms of how we think about human evolution. We show, for example, why *Graecopithecus freybergi* from the Balkan Peninsula potentially stands at the very beginning of the human evolutionary line a little more than 7 million years ago. A million years later, a still-unknown biped left its footprints on the shores of southern Greece. Its humanlike foot anatomy came as a great surprise to many scientists in the field, especially because these footprints are about 3 million years older than those left by *Australopithecus afarensis* in layers of volcanic ash at the famous site of Laetoli in Tanzania, which had previously been the oldest humanlike footprints ever discovered.

Even more startling is the recent discovery of *Danuvius guggenmosi* in Germany, which fundamentally shakes the current theory of early human evolution. Here is an ape that was partially engaging in bipedalism as it walked along branches in trees as far back as 11.6 million years ago. This challenges previous assumptions about the timing and geography of bipedal evolution. Above all, *Danuvius* suggests that bipedalism may have deep roots in the early evolution of great apes, and this changes how we look at today's chimpanzees and other living great apes. Instead of seeing them as a model for ancestors from which humans evolved, we can now view them as having undergone their own evolution, one that led to the highly specialized species we see today, with an anatomy that makes them perfectly adapted to their modern habitats. The last common ancestor of humans and chimps did not look like a chimp, but it might have looked like *Danuvius*! Europe 14 million to 6 million years ago must have been a giant laboratory where great apes were making huge evolutionary leaps forward.

All these adventures in evolution, from *Danuvius* to *Graecopithecus* and the unknown biped from Greece, have likely been driven by changes in climate and ecosystems. One major factor in this process could well have been the development of the Old World Desert Belt, which stretches from the Sahara to Mongolia. The Messinian Salinity Crisis, which happened when large parts of the Mediterranean dried out

5.5 million years ago, is part of this larger story of desertification. It was the biggest environmental crisis in the later history of Earth and had a severe impact on many species. The combined dynamic of evolving deserts, semideserts, and savannahs, amplified in the ice age that started 2.6 million years ago, may have promoted our own evolution up until recent times.

The evidence presented in this book challenges traditional assumptions about human evolution—first and foremost the idea that early evolution of humans happened exclusively in Africa. It is possible that the famous potential early hominins from Africa such as *Sahelanthropus*, *Orrorin*, and *Ardipithecus* were actually migrants from Eurasia. Paleontologists have known for a long time that the mammal fauna of today's African savannah has its evolutionary roots in Eurasia 5 million years ago, so why should early hominins be the exception?

This book covers other spectacular discoveries that suggest human evolution may be more complicated than we once thought. These early hominins include the 1.8-million-year-old *Homo georgicus* from the Caucasus, the astonishing Hobbit from Flores (*Homo floresiensis*), and *Homo luzonensis* from the Philippines. It also discusses the mysterious, over 2-million-year-old *Homo wushanensis* from China, and stone tools as old as 2.6 million years from India, China, and elsewhere in Asia. These finds suggest that human stone-tool cultures emerged contemporaneously in both Africa and Eurasia, once again questioning the idea that Africa was the sole cradle of humanity.

Perhaps the evolution of the human family starts in what some scientists have dubbed Savannahstan, the early grassland ecosystems around the Old World Desert Belt at the crossroads of continents, and not necessarily and exclusively in Africa after all.

ACKNOWLEDGMENTS

T HE RELATIONSHIPS I have had with my colleagues Nikolai Spassov and David Begun are more than just professional. We have also become friends over the many years we have known each other. Without our stimulating discussions about fossils and the talks we had while out in the field, many things would not have happened and this book would never have been written. It is a privilege for me to work with both of them, and it makes me proud that such an extraordinarily talented artist as Velizar Simeonovski has been able to represent our scientific findings with his impressive paintings.

Thanks also to my co-workers and doctoral associates and the many students who worked with unceasing curiosity and patience alongside me on digs and field explorations and generated much of the data reported in this book.

I am particularly happy to have met and learned to treasure the science journalists Rüdiger Braun and Florian Breier. Without their help and enthusiasm, I would never have dared to embark on this project. Their contributions and their many years of writing experience have made this book what it is today.

All three of us are indebted to our agent Heike Wilhelmi for her professional guidance through the publishing process and to Nadine Gibler for a wonderful collaboration in preparing the impressive illustrations. The consummate professionalism of our editors Angelica Schwab and Sara Ginolas, who oversaw the process, and the detailed comments of the copy editor, Gisela Klemt, and graphic designer, Tanja Zielezniak, were all unbelievably important for a wide-ranging book such as this one.

I would especially like to thank my wife, Nicole Preußner. She had to do without vacations and many weekends and was unfailingly understanding and patient during the long process that finally culminated in the publication of this book.

NOTES

Introduction and Part 1: El Graeco and the Split Between Chimpanzees and Humans

1. Per Donat and Ullrich Herbert, *Wie sich der Mensch aus dem Tierreich erhob* (Berlin: Kinderbuchverlag, 1972).

2. Bruno von Freyberg, "Die Pikermi-Fauna von Tour la Reine (Attika)," *Annales Geologiques des Pays Helléniques* 3 (1949): 7–10.

3. Anton Lindermayer, "Die fossilen Knochenreste in Pikermi in Griechenland," in *Correspondenz-Blatt des Zoolisch-Mineralogischen Vereines in Regensburg* 14 (Regensburg: F. Pustet, 1860), 109.

4. Andreas Johann Wagner, "Fossile Ueberreste von einem Affenschädel und anderen Säugethierreste aus Griechenland," *Gelehrte Anzeigen von dem Bayerisches Academie der Wissenschaften in München* 38 (1839): 301–312.

5. Ottenio Abel, *In der Buschsteppe von Pikermi in Attica* (Jena: Fischer, 1922), 75–165.

6. Albert Gaudry, *Animaux fossiles et géologie de l'Attique* (Paris: F. Savy, 1862–1867).

7. Bruno von Freyberg, *Im Banne der Erdgeschichte* (Erlangen: Junge & Sohn, 1977).

8. This dispute is mentioned in a communication dated August 4, 1994, from Dr. D. P. Andrews, Natural History Museum, London, to Professor G. Nollau, University of Erlangen. Personal collection of the author.

9. Jochen Fuss, Nicolai Spassov, David R. Begun, and Madelaine Böhme, "Potential Hominin Affinities of *Graecopithecus* From the Late Miocene of Europe," *PLOS ONE* 12, no. 5 (May 22, 2017): e0177127.

10. The period of time when the human evolutionary line likely split from chimpanzees—a split that has been estimated based on genetic analysis—spans from 13 million to 6 million years before today.

Part 2: The Real Planet of the Apes

1. Lartet was a student of the grand master of comparative anatomy and founder of paleontology, Georges Cuvier (1769–1832). In 1812, Cuvier insisted there was no such thing as a fossil human or a fossil primate. Scientists were collecting many fossils at the time, but none from primates. Cuvier was unaware he himself had collected a lemur from a limestone quarry in Paris. He called his find *Adapis parisiensis* and thought it was a primitive hoofed animal. Lartet was the first person to describe a fossil primate recognized as such (*Pliopithecus*) as well as *Dryopithecus fontani*, the first fossil great ape.

2. The cave bear was a widely distributed species of bear in the ice age, and its remains were found in caves throughout Europe.

3. Only later did people realize that Neanderthal fossils had been collected in Belgium in 1829 and in Gibraltar in 1848, but at the time they were not recognized as belonging to a new species.

4. He is referring to the Miocene, a geological epoch between 23 million and 5.3 million years before the present.

5. Charles Darwin, *The Descent of Man, and Selection in Relation to Sex* (London: John Murray, 1871).

6. The lower jawbone is about 600,000 years old and is the type specimen for the species.

7. Mirjana Roksandic, Predrag Radović, and Joshua Lindal, "Revising the Hypodigm of *Homo heidelbergensis*: A View From the Eastern Mediterranean," *Quaternary International* 466 (2018): 66–81.

8. Fossils from juveniles are particularly difficult to interpret because they are often distinctly different from mature individuals.

9. Later calculations for the Taung Child came up with even smaller values, and other finds of *Australopithecus africanus* yielded slightly higher values. Ralph L. Holloway, "Australopithecine Endocast (Taung Specimen, 1924): A New Volume Determination," *Science* 16, no. 3924 (May 22, 1970): 966–968. D. Falk and R. Clarke, "Brief Communication: New Reconstruction

of the Taung Endocast," *American Journal of Physical Anthropology* 134, no. 4 (December 2007): 529–534.

10. That was when University of Oxford anatomist Wilfrid Le Gros Clark acknowledged in a paper published in *Nature* that *Australopithecus africanus* was an early hominin. A few years later, Le Gros Clark was involved in exposing the Piltdown Man as a hoax.

11. The 1.75-million-year-old lower jawbone from the Olduvai Gorge, sometimes referred to as Johnny's Child, is the type specimen for *Homo habilis*. See Louis Leakey, Phillip V. Tobias, and John Russell Napier, "A New Species of the Genus *Homo* From Olduvai Gorge," *Nature* 202 (April 4, 1964): 7–9.

12. To this day, tools like this are referred to as belonging to the Oldowan Culture. See M. D. Leakey, "A Review of the Oldowan Culture From Olduvai Gorge, Tanzania," *Nature* 210 (April 30, 1966): 462–466.

13. This classification remains controversial. Some experts argue that great apes and even birds also use tools. Others claim that *Homo habilis* bears a closer resemblance to the australopithecenes than to *Homo erectus*. Louis Leakey's son Richard Leakey later found more *Homo habilis* fossils at Lake Turkana in Kenya. Richard Leakey's wife, Maeve, and daughter, Louise, both paleoanthropologists as well, described more finds from this area in 2007. (F. Spoor, M. Leakey, P. Gathogo, et al., "Implications of New Early *Homo* Fossils From Ileret, East of Lake Turkana, Kenya," *Nature* 448 [August 9, 2007]: 688–691.) All the finds accepted as *Homo habilis* fall into the time period between 1.75 million and 1.44 million years ago. That proves that *Homo habilis* existed at the same time as *Homo erectus*. For this reason, there is some dispute about whether *Homo erectus* evolved from *Homo habilis* or whether another species is the direct ancestor of *Homo erectus*.

14. Even today it is still not clear if Lucy walked upright all the time or only occasionally, perhaps spending a lot of time in trees. Some researchers believe she died when she fell out of a tree. See John Kapenman, Richard Ketcham, Stephen Pearce, et al., "Perimortem Fractures in Lucy Suggest Mortality From Fall Out of Tall Tree," *Nature* 537 (August 29, 2016): 503–507.

15. African finds that are similar to *Homo erectus* are often given the species name *Homo ergaster* to distinguish them from the finds in Asia.

16. Molecular biology allows us to estimate when two species split from ancestors they shared in common. The more mutations there are—that is to say, changes in the pattern of specific sections of DNA—the longer separate

evolutionary paths have been followed. The most difficult calculation is to determine the rate at which mutations are taking place—in other words, to estimate how fast the molecular clock is running.

17. Wolfram Kürschner, Zlatkko Kvaček, and David L. Dilcher, "The Impact of Miocene Atmospheric Carbon Dioxide Fluctuations on Climate and the Evolution of Terrestrial Ecosystems," PNAS (January 15, 2008): 449–453.

18. The causes of warming in the Miocene are not yet fully explained. The most likely reasons are a change in the pattern of ocean currents and an elevated level of carbon dioxide in the air after a great deal of volcanic activity.

19. David R. Begun, The Real Planet of the Apes (Princeton: Princeton University Press, 2016).

20. Madelaine Böhme, Angela A. Bruch, and Alfred Selmeier, "The Reconstruction of the Early and Middle Miocene Climate and Vegetation in the North Alpine Foreland Basin as Determined From the Fossil Wood Flora," Palaeogeography, Palaeoclimatology, Palaeoecology 195 (2003): 91–114.

21. Madelaine Böhme, "Miocene Climatic Optimum: Evidence From Lower Vertebrates of Central Europe," Palaeogeography, Palaeoclimatology, Palaeoecology 195 (2003): 389–401.

22. Madelaine Böhme, "Migration History of Air-Breathing Fishes Reveal Neogene Atmospheric Circulation Pattern," Geology 32 (2004) 393–396.

23. Primates (monkeys, apes, and prosimians) arose in the northern continents of North America and Asia more than 60 million years ago, then some of them migrated to Africa.

24. Embrithopods are herbivores about the size of a rhinoceros with two parallel bony horns. These primeval mammals may have behaved like hoofed animals but they are actually distantly related to elephants. They made it as far as some of the islands in the Tethys Ocean, but they never made it all the way across to Europe.

25. Christian-Dietrich Schönweise and Christoph Buchal, Klima: Die Erde und Ihre Atmosphäre in Wandel der Zeiten (Helmholtz: Cologne, 2010).

26. Shiming Wan, Wolfram M. Kürschner, Peter D. Clift, et al., "Extreme Weathering/Erosion During the Miocene Climatic Optimum: Evidence From Sediment Record in the South China Sea," Geophysical Research Letters 36, no. 19 (October 2009).

27. David R. Begun, The Real Planet of the Apes (Princeton: Princeton University Press, 2016).

28. Isaac Casanovas-Vilara, Anneke Madern, David M. Albaa, et al., "The Miocene Mammal Record of the Vallès-Penedès Basin (Catalonia)," *Comptes Rendus Palevol* 15 (2016): 791–812.

29. The site also yielded the oldest evidence of a fossil gibbon ever found.

30. At 45 degrees of latitude, the days around the summer solstice last about fifteen hours, whereas around the winter solstice, they last only nine. This is an enormous difference when compared with the constant equal split between day and night at the equator.

31. James T. Kratzer, Miguel A. Lanaspa, Michael N. Murphy, et al., "Evolutionary History and Metabolic Insights of Ancient Mammalian Uricases," *PNAS* 111, no. 10 (March 11, 2014): 3763–3768.

32. Jochen Fuss, Gregor Uhlig, and Madelaine Böhme, "Earliest Evidence of Caries Lesion in Hominids Reveal Sugar-Rich Diet for a Middle Miocene Dryopithecine From Europe," *PLOS ONE* 13, no. 8 (August 13, 2018): e0203307.

33 Richard Johnson and Peter Andrews, "Fructose, Uricase, and the Back-to-Africa Hypothesis," *Evolutionary Anthropology* 19, no. 6 (2010): 250–257.

34. When we talk of Homininae (hominines), we define them as African apes and humans and their fossil ancestors. This is confusing if you look at their evolutionary history, because at this phase of their development, the only finds we have are from outside Africa. The terminology in this case reflects only their current distribution.

35. Madelaine Böhme, August Ilg, and Michael Winklhofer, "Late Miocene 'Washhouse' Climate in Europe," *Earth and Planetary Science Letters* 275 (2008): 393–401.

36. David R. Begun, *The Real Planet of the Apes* (Princeton: Princeton University Press: 2016).

37. Sigulf Guggenmos died on September 15, 2018. Unfortunately, he did not live to see the publication of the original German version of this book or the first published article on the apes of Hammerschmiede.

38. Helmut Mayr and Volker Fahlbusch, "Eine unterpliozäne Kleinsäugerfauna aus der Oberen Süßwasser-Molasse Bayerns," *Mitteilungen der Bayerischen Staatssammlung für Paläontologie und historische Geologie* 15 (1975): 91–111.

39. We used magnetostratigraphy to date the layer where we found the fossils (see Chapter 5). To do that, we drilled down nearly 500 feet (150 meters) into the ground near Irsee Abbey. As the hole extended beyond the clay pit

at the base of the mountain, we could determine and compare the paleo-
magnetic signals from the drilling core with those from the rocks in the pit.
This method dated the layer where the fossils were found to 11.62 million
years ago. See Uwe Kirscher, Jerome Prieto, Valerian Bachtadse, et al., "A
Biochronologic Tie-Point for the Base of the Tortonian Stage in European
Terrestrial Settings: Magnetostratigraphy of the Topmost Upper Fresh-
water Molasse Sediments of the North Alpine Foreland Basin in Bavaria
(Germany)," *Newsletters on Stratigraphy* 49, no. 3 (August 2016): 445-467.

40. The species *Miotragocerus monacensis* was first described in 1928 by Max
Schlosser from deposits along the banks of the Isar River in Munich.

41. Members of the genus *Miotragocerus* probably had two pointed horns like
pronghorns today.

42. *Tetralophodon longirostris* is a gomphothere (a trunked animal with tusks
on both its upper and lower jaws), which in German are called "teat-
toothed" thanks to the pronounced rounded bumps on their teeth. We
found bones from both old and newborn individuals of this species at
the dig in Hammerschmiede.

43. Today, muntjaks live mostly in dense mountain forests in south and southeast Asia. The bones from the dig at Hammerschmiede probably belong to
an unidentified species.

44. *Hoploaceratherium belvederense* is a relatively long-legged, hornless rhinoceros. It was first described by Wang in the Belvedere beds in the Vienna
basin.

45. Pheasants in the genus *Miophasaneus* were the most common gallinaceous
birds during the Miocene (23 million to 5 million years ago) in Europe.

46. The prolific vegetation in and around these ponds formed the coal beds
found in the area around Irsee Abbey.

47. Given the prevailing westerly winds in the Allgäu, we have to assume most
of the rain fell in the summer. Thanks to the proximity of the Alps, it likely
rained often and heavily during heat-induced thunderstorms at that time
of year.

48. The snapping or alligator turtle, *Chelydropsis murchisoni*, was the most
common of a total of five different species of turtle in Hammerschmiede.
Like their living relatives in North America, these turtles, which were up to
30 inches (80 centimeters) long, were probably voracious predators.

49. The bear-dog *Amphicyon* belongs to an extinct family of predators (Amphi-
cyonidae). It is not particularly closely related to either dogs or bears. It
was a swift-footed predator, and unlike the catlike predators, it crushed

and ate the bones of its prey. The species *Amphicyon major*, which we found at Hammerschmiede, was the last and largest species of bear-dog that ever lived. They grew as large as modern lions and died out about 10 million years ago, probably because they could not compete with saber-toothed cats and large hyenas.

50. We found two different species of pigs at Hammerschmiede. The more common one, *Parachleuastochoerus steinheimensis*, is relatively closely related to European wild boar, but with a maximum body length of 4 feet (1.2 meters) it was not nearly as large.

51. Hyenas of the genus *Miohyaena* lived near Hammerschmiede. They ate carrion and bones, as modern hyenas do, but they were slightly smaller than hyenas are today.

52. We did not find any gravel from the Alps in the stream sediments, but we did find sand and microfossils from what is known as the Upper Marine Molasse. These marine deposits can be found just a few miles south of Hammerschmiede in the mountain ridges, where alpine tectonics thrust the basin upward. Today, the river Günz starts here, unlike the rivers Iller and Wertach, which start in the Alps.

53. The ones found in Hammerschmiede probably belong to a new species in the genus *Ancylotherium*.

54. Forest horses belong to their own subfamily (Anchitheriinae) in the horse family. They differ from true horses in that they have three functional toes on their fore and hind legs. Their shorter side toes, however, made contact with the ground only when they were standing still. When they walked, they relied on their powerful center hoofs. Large anchitherians are typically found in North America. They are definitely rare in Eurasia, which they reached by crossing the Bering Land Bridge. The finds in Hammerschmiede are probably of a newly discovered species.

55. Beech leaves are the most common of the rare plant finds from Hammerschmiede.

56. The oldest-known panda bear, the species *Kretzoiarctos beatrix*, was relatively common in the foothills of the Alps over 11 million years ago. It was smaller than its living relative and probably did not eat bamboo, but it was a vegetarian.

57. *Albanensia albanensis*, 3 feet (1 meter) wide with its limbs stretched to their fullest extent, was one of the largest flying squirrels. It is one of three species of flying squirrel found at Hammerschmiede. Like all squirrels, *Albanensia* ate a variety of nuts, including beech nuts.

58. These facts can be deduced from the extent and composition of the sediment that filled the channel. After heavy rain, the river certainly grew to be about 80 feet (25 meters) wide.

59. The giant salamander *Andras scheuchzeri* is an important fossil in the history of science, as skeletons from this species were used in 1726 by the Zurich doctor Johann Jacob Scheuchzer as proof of Noah's flood.

60. Catfish of the genus *Silurus* are the most abundant animal fossils in Hammerschmiede. We have found well over two hundred individuals. They grow no more than 8 inches (20 centimeters) long and belong to a new species. They are the oldest true catfish in the world.

61. *Pseudaelurus quadridentatus* was a true cat. The much larger saber-toothed cats belong to a different family. They arrived in Europe a few hundred thousand years after the sediments at Hammerschmiede were laid down.

62. Udo's prominent cheekbones were mostly hollow. Such pronounced sinus cavities function as resonance chambers, increasing the volume of the voice and allowing it to travel long distances. It is therefore likely that Udo was very loud.

63. These are the genera *Rudapithecus* (10 million years old) from Hungary and *Hispanopithecus* (9.6 million years old) from Spain.

64. Calculations about the body mass of extinct apes are made using a fixed relationship between weight and the size of the bones that support the body mass, for example, the thighbone. We have thighbones for all three adult *Danuvius* individuals and the weight calculations are based on these.

65. This has to do with adaptations to moving by swinging from branches, where all the weight is carried on one arm. Longer arms have greater leverage, which allows for quicker movement through the branches, but centrifugal forces greatly increase as bodies get heavier. Gorillas are large, hardly swing by their arms at all, and their arms are "only" 30 percent longer than their legs. Orangutans are moderately heavy and their arms are up to 50 percent longer. Lesser apes—that is to say, gibbons—are extremely light and their arms are up to 80 percent longer.

66. The big toes of monkeys and apes that spend most of their time on the ground, such as baboons or mountain gorillas, tend to be turned inward. This allows the soles of their feet to make better contact with the ground.

67. C. Owen Lovejoy and Melanie A. McCollum, "Spinopelvic Pathways to Bipedality: Why No Hominids Ever Relied on a Bent-Hip–Bent-Knee Gait," *Philosophical Transactions of the Royal Society B* 365 (October 27, 2010): 3289–3299.

68. Tropical figs are the mainstay of chimpanzee diets today. These fruits, however, grow on trunks (so-called cauliflory) or on very thick branches (ramiflory). From the perspective of nutrition, therefore, what is of most interest to chimpanzees is the central part of the fig tree: the trunk and the main branches. They are less interested in the outermost branches high in the canopy. When they scale a new tree, they climb up the main trunk. They usually don't want to go any farther. In contrast, there is the Carinthian great ape, *Dryopithecus carinthiacus*, who, we know, particularly liked to eat fruits such as cherries, plums, and grapes. These grow at the ends of thin branches, where they are hard to reach, or on vines. See Jochen Fuss, Gregor Uhlig, and Madelaine Böhme, "Earliest Evidence of Caries Lesion in Hominids Reveal Sugar-Rich Diet for a Middle Miocene Dryopithecine From Europe," *PLOS ONE* 13, no. 8 (August 30, 2018): e0203307.

Part 3: The Cradle of Humanity: Africa or Europe?

1. Jochen Fuss, Nicolai Spassov, David R. Degun, and Madelaine Böhme, "Potential Hominin Affinities of *Graecopithecus* From the Late Miocene of Europe," *PLOS ONE* 12, no. 5 (2107): e0177127.

2. Chimpanzees wade through water on two feet. Gibbons can walk on two legs for short distances between trees (their arms are just too long). Male gorillas stand up on two legs when they want to look imposing. Orangutans can balance on branches using two legs. A number of monkeys "stand up" to reach for branches.

3. Jack T. Stern, "Climbing to the Top: A Personal Memoir of *Australopithecus afarensis*," *Evolutionary Anthropology* 9, no. 3 (June 16, 2000): 113-133.

4. A few early hominins from Africa also show a rudimentary arch. *Australopithecus afarensis*, however, clearly shows that there was some variability in how the arch was formed.

5. These plankton were foraminifera, single-cell marine organisms, most smaller than 0.02 inches (0.5 millimeters), that lived in chambered chalky shells. Foraminifera are important fossils for earth scientists who do oceanographic dating. The scientists use them to calculate the age of sediments.

6. Jeremy M. DeSilva and Zachary J. Throckmorton, "Lucy's Flat Feet: The Relationship Between the Ankle and Rearfoot Arching in Early Hominins," *PLOS ONE* 5, no. 12 (December 28, 2010): e14432.

7. Kevin E. Langergraber, Kay Prüfer, Carolyn Rowney, et al., "Generation Times in Wild Chimpanzees and Gorillas Suggest Earlier Divergence Times in Great Ape and Human Evolution," *PNAS* 109, no. 39 (September

25, 2012): 15716–15721. And, Qiaomei Fu, Heng Li, Priya Moorjani, et al., "Genome Sequence of a 45,000-Year-Old Modern Human From Western Siberia," *Nature* 514 (October 23, 2014): 445–449.

8. Jim Giles, "The Dustiest Place on Earth," *Nature* 434, no. 12 (April 2005): 816–819. And, Charlie S. Bristow, Nick Drake, and Simon Armitage, "Deflation in the Dustiest Place on Earth: The Bodélé Depression, Chad," *Geomorphology* 105, no. 1 (2009): 50–58.

9. In 1995, the Mission Paléontologique Franco-Tchadienne discovered the lower jawbone and one tooth from a 3.5-million-year-old early hominin at the Koro Toro site in Chad. The researchers named it *Australopithecus bahrelghazali*, aka Abel. Many other researchers, however, thought the fossils were the remains of a local variant of *Australopithecus afarensis*.

10. Alain Beauvilain has published partial accounts of the discovery of Toumaï in many different works, including Alain Beauvilain and Yves Le Guellec, "Further Details Concerning Fossils Attributed to *Sahelanthropus tchadensis* (Toumaï)," *South African Journal of Science* 100, no. 3–4 (March 2004): 142–144.

11. Ann Gibbons, *The First Human* (New York: Random House, 2006).

12. Michel Brunet, Franck Guy, David Pilbeam, et al., "A New Hominid From the Upper Miocene of Chad, Central Africa," *Nature* 418 (2002): 145–151.

13. Milford H. Wolpoff, Brigitte Senut, Martin Pickford, and John Hawks, "*Sahelanthropus* or '*Sahelpithecus*'?" *Nature* 419 (2002): 581–582.

14. Christoph P. E. Zollikofer, Marcia S. Ponce de León, Daniel E. Lieberman et al., "Virtual Cranial Reconstruction of *Sahelanthropus tchadensis*," *Nature* 434 (2005): 755–759.

15. It almost never happens that a fossil human or great ape skull retains all the area around the brain. Usually these parts of the skull are bitten through around the time of death by carnivores (for example, hyenas) that want to get to the nutritious brain inside. In those cases, all that is left is the facial portion of the skull. The conditions required for preserving the back of the skull improved when Neanderthals and modern humans instituted burial rites, because burials make it more difficult for scavengers to get to the bodies.

16. Milford H. Wolpoff, John Hawks, Brigitte Senut, et al., "An Ape or the Ape: Is the Toumaï Cranium TM 266 a Hominid?" *PaleoAnthropology* (2006): 36–50.

17. Martin Pickford and Roberto Macchiarelli, personal communication, July 2010.

18. Brian G. Richmond and William L. Jungers, "*Orrorin tugenensis* Femoral Morphology and the Evolution of Human Bipedalism," *Science* 319 (March 21, 2008): 1662–1665.

19. Ewen Callaway, "Controversial Femur Could Belong to Ancient Human Relative," *Nature* 553 (January 22, 2018): 391–392.

20. Roberto Macchiarelli in a letter to the scientific community, October 30, 2017.

21. Roberto Macchiarelli, "Premiers hominines, premiers humains : des problèmes, plusieurs questions, des prospectives," *Prospectives du CNRS-INEE* (2017).

22. Alain Beauvilain, "The Contexts of Discovery of *Australopithecus bahrelghazali* (Abel) and of *Sahelanthropus tchadensis* (Toumaï): Unearthed, Embedded in Sandstone, or Surface Collected?" *South African Journal of Science* 104 (May/June 2008): 165–168.

23. Alain Beauvilain and Jean-Pierre Watté, "Toumaï (*Sahelanthropus tchadensis*) a-t-il été inhumé?" *Bulletin de la Société Géologique de Normandie et des Amis du Muséum du Havre* 96 (2009): 19–26.

24. Mathieu Schuster, Philippe Duringer, Jean-François Ghienne, et al., "The Age of the Sahara Desert," *Science* 311, no. 5762 (February 10, 2006): 821.

25. William H. Kimbel and Lucas K. Delezene, "'Lucy' Redux: A Review of Research on *Australopithecus afarensis*," *Yearbook of Physical Anthropology* 140, suppl. 49 (January 1, 2009): 2–48.

26. Jack T. Stern, "Climbing to the Top: A Personal Memoir of *Australopithecus afarensis*," *Evolutionary Anthropology* 9, no. 3 (June 16, 2000): 113–133.

27. C. Owen Lovejoy, "The Natural History of Human Gait and Posture Part 2, Hip and Thigh," *Gait Posture* 21, no. 1 (January 2005): 113–124.

28. This is the gist of what is written in Jean-Baptiste de Lamarck, *Philosophie zoologique, ou, Exposition des considérations relatives à l'histoire naturelle des animaux* (Paris: Musée d'Histoire Naturelle, 1809).

29. M. Domínguez-Rodrigo, "Is the 'Savanna' Hypothesis a Dead Concept for Explaining the Emergence of the Earliest Hominins?" *Current Anthropology* 55, no. 1 (February 2014): 59–81.

30. Fred Spoor, Philipp Gunz, Simon Neubauer, et al., "Reconstructed *Homo habilis* Type OH 7 Suggests Deep-Rooted Species Diversity in Early *Homo*," *Nature* 519 (March 5, 2015): 83–86.

31. At the present time, the classification of these early species from Africa, *Homo rudolfensis*, *Homo habilis*, and *Homo ergaster*, is being disputed.

32. The species *Homo habilis* and *Homo rudolfensis* share many features with members of the genus *Australopithecus*. A few researchers, therefore, consider them to belong to this genus and regard them as early hominins and not as species of early humans.

33. Scientists recently identified a 2.44-million-year-old Oldowan culture in Ain Bouchert, Algeria, 62 miles (100 kilometers) from the Mediterranean coast.

34. Claire Gaillard, Mukesh Singh, Anne Dambricurt Malassé, et al., "The Lithic Industries on the Fossiliferous Outcrops of the Late Pliocene Masol Formation, Siwalike Frontal Range, Northwestern India (Punjab)," *Comptes Rendus Palevol* 15, nos. 3–4 (February/March 2016): 341–357.

35. Guangbiao Wei, Wanbo Huang, Shaokun Chen, et al., "Paleolithic Culture of Longgupo and Its Creators," *Quaternary International* 354 (December 15, 2014): 154–161.

36. *Gigantopithecus blacki* resided throughout Southeast Asia. It is known from about a thousand teeth and a few lower jaw fragments from ten different caves in China, Vietnam, and Thailand. With a height of 6 feet 6 inches (2 meters), which is an extremely rough estimate, it is the largest ape that ever lived.

37. Huang Wanpo, Russell Ciochon, Gu Yumin, et al., "Early *Homo* and Associated Artefacts From Asia," *Nature* 378 (November 16, 1995): 275–278.

38. Russell L. Ciochon, "The Mystery Ape of Pleistocene Asia," *Nature* 459 (June 11, 2009): 910–911.

39. Yahdi Zaim, Russell L. Ciochon, Joshua M. Polanski, et al., "New 1.5 Million-Year-Old *Homo erectus* Maxilla From Sangiran (Central Java, Indonesia)," *Journal of Human Evolution* 61, no. 4 (July 2011): 363–376.

40. Rixiang X. Zhu, Richard Potts, Yongxin X. Pan, et al., "Early Evidence of the Genus *Homo* in East Asia," *Journal of Human Evolution* 55 (2008): 1075–1085.

41. Zhao-Yu Zhu, Robin Dennell, Wei-Wen Huang, et al., "New Dating of the *Homo erectus* Cranium From Lantian (Gongwangling), China," *Journal of Human Evolution* 78 (2015): 144–157.

42. Yonas Beyene, Shigehiro Katoh, Giday Wolde Gabriel, et al., "The Characteristics and Chronology of the Earliest Acheulean at Konso, Ethiopia," *PNAS* (January 29, 2013): 1584–1591.

43. V. E. Shchelinsky, Maria Gurova, Alexey S. Tesakov, et al., "The Early Pleistocene Site of Kermek in Western Ciscaucasia (Southern Russia): Stratigraphy, Biotic Record and Lithic Industry (Preliminary Results)," *Quaternary International* 393 (January 14, 2016): 51–69.

44. Franz Weidenreich, "The Skull of Sinanthropus pekinensis: A Comparative Study on a Primitive Hominid Skull," *Palaeontologia Sinica* 127 (1943).

Part 4: Climate Change as a Driver of Evolution

1. Paleogenetics is particularly important, especially for the later chapters in the story of human evolution. The oldest human DNA that researchers have been able to decode so far is about 400,000 years old and comes from fossils that were discovered in a cave in northern Spain called the Sima de los Huesos ("the chasm of the bones"). The oldest DNA ever discovered comes from the bone of a horse that researchers found in the permafrost in Yukon, Canada.

2. The last of these animals died in captivity. Aurochs and Tasmanian tigers died out much earlier in the wild.

3. The American physical chemist Willard Frank Libby developed this well-known method back in the late 1940s. In 1960 he was awarded the Nobel Prize in Chemistry for his discovery.

4. A tachymeter is an instrument used for taking quick measurements of geological data. It measures not only vertical and horizontal angles but also distances.

5. Heinrich Schliemann (1822–1890) was a German archeologist and is credited with discovering ancient Troy. While searching for the treasure of Priam, he had deep trenches dug through the hills around the city, which meant much important information was irretrievably lost. After being criticized for how he had conducted the dig, he greatly improved his methods. Today, therefore, he is considered one of the founders of modern archeology.

6. It is no accident that the Acropolis stands where it does on a limestone massif high above the basin of Athens. Tectonic events over 30 million years ago thrust a massive covering of limestone over an impermeable layer of slate. However, apart from a few relic peaks, including the Acropolis massif, the limestone layer has eroded away. The numerous springs at the base of the massif have been highly prized since the Neolithic Age because they provide fresh water all year long. The Clepsydra Spring, which flows to the surface in a cave, is a particularly good place to see where the porous limestone, which collects rainwater like a sponge and gradually releases it, and the impervious slate meet.

7. *Pliohyrax graecus.*

8. Asier Larramendi, "Shoulder, Height, Body Mass, and Shape of Proboscideans," *Acta Palaeontologica Polonica* 61, no. 3 (2016): 537–574.

9. Ottenio Abel, *In der Buschsteppe vor Pikermi in Attica* (Jena: Fischer, 1922), 75–165.

10. Homer, *The Iliad*, trans. T. S. Norgate (Edinburgh: Williams & Norgate, 1864), 282.

11. A. Goudie and N. J. Middleton, *Desert Dust in the Global System* (Berlin: Springer, 2006).

12. The reddish-orange color of the dust is due to the high proportion of iron oxide and iron hydroxide it contains.

13. There are more bromide ions in salt in the ocean than in salt formed when water dries up on land, where bromide is present in only very small quantities.

14. Jef Vandenberghe, "Grain Size of Fine-Grained Windblown Sediment: A Powerful Proxy for Process Identification," *Earth-Science Reviews* 121 (June 2103): 18–30.

15. We also proved the presence of Saharan sand in layers of a similar age in southern France.

16. In areas with minimal rainfall, the wind is the main sculptor of the landscape. You could say deserts are constantly being blown away because there is no vegetation to hold the sand and dust. Unprotected, they are at the mercy of the wind, and today 2,000 tons of rock dust are distributed around the world in dust storms every year. Y. Shao, "Dust Cycle: An Emerging Core Theme in Earth System Science," *Aeolian Research* 2, no. 4 (March 2011): 181–204.

17. These estimates are based on ocean temperatures calculated for the Mediterranean at that time. Alexandrina Tsanova, Timoth Herbert, and Laura Peterson, "Cooling Mediterranean Sea Surface Temperatures During the Late Miocene Provide a Climate Context for Evolutionary Transitions in Africa and Eurasia," *Earth and Planetary Science Letters* 419 (June 1, 2015): 71–80.

18. Organic material such as leaves and pollen can quickly decompose or oxidize.

19. Based on the statistics from the phytoliths and chemical analysis of the soils, we estimated trees covered about 40 percent of the Basin.

20. The scientific names for these subfamilies of grasses are *Panicoideae* and *Chloridoideae*.

21. Albert Gaudry, *Animaux fossiles et géologie de l'Attique* (Paris: F. Savy, 1862–1867).

22. Martín Iriondo and Daniela Kröhling, "Non-Classical Types of Loess," *Sedimentary Geology* 202, no. 3 (December 1, 2007): 352–368.

23. Julia F. Morton, "Cattails (*Typha* spp.)—Weed Problem or Potential Crop," *Economic Botany* 29, no. 1 (January-March 1975): 7–29.

24. S. M. Plaisted, "The Edible, Incredible Cattail," *Wild Food: Proceedings of the Oxford Symposium on Food and Cookery* (Oxford: Oxford Symposium, 2004), 260–262.

25. Steffen Guido Fleischhauer, Jürgen Guthmann, and Roland Spiegelberger, *Enzyklopädie essbarer Wildpflanze* (Aarau: AT Verlag, 2016).

26. Richard Wrangham, *Catching Fire: How Cooking Made Us Human* (New York: Basic Books, 2009).

27. The mammals you find in the Sahara include the fennec fox, the addax or screwhorn antelope, and a tiny jumping rodent called the desert jerboa.

28. There are two species of camel, the one-humped dromedary that is mostly found from North Africa to India, and the two-humped Bactrian camel, which is found from western Asia to China. However, camels originated in North America. No one yet knows exactly when and how they reached Eurasia, but they certainly made the journey more than 6 million years ago. They probably used land bridges in the North Pacific that predated the Bering Land Bridge.

29. Faisal Almathen, Pauline Charruau, Elmira Mohandesan, et al., "Ancient and Modern DNA Reveal Dynamics of Domestication and Cross-Continental Dispersal of the Dromedary," *PNAS* 113, no. 24 (June 14, 2016): 6707–6712.

30. The modern-day Tuareg trace their culture back to these first nomadic arrivals in the Sahara.

31. When we look back today, the relatively small settled area along the narrow, fertile corridor in the middle of an enormous desert seems tiny and cramped, but the mighty Nile River gave the Egyptians of that time all they needed. The Nile gets its water from the damp monsoon regions of equatorial Africa, where it originates in the mountainous regions of Rwanda and Burundi. Heavy rainfall in the upper stretches of the river leads to regular flooding in the middle and lower stretches. Since time immemorial, the Nile has been creating fertile bottomland beyond its banks that continues to be the foundation for agriculture in Egypt.

32. Georgi N. Markov, "The Turolian Proboscideans (Mammalia) of Europe: Preliminary Observations," *Historia Naturalis Bulgarica* 19 (2008): 153–178.

33. Madelaine Böhme, Nikolai Spassov, Martin Ebner, et al., "Messinian Age and Savannah Environment of the Possible Hominin *Graecopithecus* From Europe," PLOS ONE 12, no. 5 (May 22, 2017): e0177347.

34. Madelaine Böhme, Christiaan Van Baak, Jerome Prieto, et al., "Late Miocene Stratigraphy, Palaeoclimate and Evolution of the Sandanski Basin (Bulgaria) and the Chronology of the Pikermian Faunal Changes," *Global and Planetary Change* 170 (2018): 1–19.

35. Elephant shrews get their name because they look like shrews but are more closely related to elephants. See Christophe J. Douady, François Catzeflis, Jaishree Raman, et al., "The Sahara as a Vicariant Agent, and the Role of Miocene Climatic Events, in the Diversification of the Mammalian Order Macroscelidea (Elephant Shrews)," PNAS 100, no. 14 (June 23, 2003): 8325–8330.

36. Syed Shujait Ali, Martin Pfosser, Wolfgang Wetschnig, et al., "Out of Africa: Miocene Dispersal, Vicariance, and Extinction Within Hyacinthaceae Subfamily Urgineoideae," *Journal of Integrative Plant Biology* 55, no. 10 (October 2013): 950–964.

37. At the beginning of the last African wet phase, eleven thousand years ago, the first settled agricultural cultures developed in the Fertile Crescent, an area that stretches from Israel through Syria to Iran. This was the beginning of the Neolithic Revolution: the discovery of agriculture and animal husbandry. When this wet phase ended, this lifestyle also ended in the Sahara, and a few highly developed cultures in Mesopotamia and Central Asia collapsed.

38. Hans-Joachim Pachur and Norbert Altmann, *Die Ostsahara im Spätquartär: Ökosystemwandel im größten hyperariden Raum der Erde* (Berlin: Springer, 2006).

39. Lawrence J. Flynn, Alisa J. Winkler, Margarita Erbaeva, et al., "The Leporid Datum: A Late Miocene Biotic Marker," *Mammal Review* 44, nos. 3/4 (2015): 164–176.

40. Andossa Likius, Patrick Vignaud, and Michel Brunet, "A New Species of *Bohlinia* (Mammalia, Giraffidae) From the Late Miocene of Toros-Menalla," *Comptes Rendus Paleovol* 6, no. 3 (March 2007): 211–220.

41. Faysal Bibi, "Mio-Pliocene Faunal Exchanges and African Biogeography: The Record of Fossil Bovids," PLOS ONE 6, no. 2 (February 16, 2011): e16688.

42. The Messinian age is the final or uppermost age of the Miocene. It lasted from 7.2 million to 5.3 million years ago.

43. Kenneth J. Hsü, *The Mediterranean Was a Desert* (Princeton: Princeton University Press, 1983).

44. Marco Roveri, Rachel Flecker, Wout Krijgsman, et al., "The Messinian Salinity Crisis: Past and Future of a Great Challenge for Marine Sciences," *Nature* 400 (1999): 652–655.

45. Wout Krijgsman, Frits Hilgen, Isabella Raffi, et al., "Chronology, Causes, and Progression of the Messinian Salinity Crisis," *Nature* 400 (August 12, 1999): 652–655.

46. P. Th. Meijer and Wout Krijgsman, "Quantitative Analysis of the Desiccation and Re-Filling of the Mediterranean During the Messinian Salinity Crisis," *Earth and Planetary Science Letters* 240, no. 2 (December 1, 2005): 510–520.

47. The word salt here includes not only chlorides, such as table salt, but also sulfates, such as gypsum.

48. Gypsum (calcium sulfate) is more soluble than carbonates such as limestone and calcium carbonate. Therefore, when brine evaporates, the first salts to be deposited are always gypsum or rock salt, and then, occasionally, potash salt (potassium chloride).

49. Rob Govers, "Choking the Mediterranean to Dehydration: The Messinian Salinity Crisis," *Geology* 37, no. 2 (February 1, 2009): 167–170.

50. Adele Bertini, "The Northern Apennines Palynological Record as a Contribute for the Reconstruction of the Messinian Paleoenvironments," *Sedimentary Geology* 188–189 (June 25, 2006): 235–258.

51. Lisa N. Murphy, "The Climate Impact of the Messinian Salinity Crisis" (dissertation, University of Maryland, 2010).

52. William B. F. Ryan, "Decoding the Mediterranean Salinity Crisis," *Sedimentology* 56, no. 1 (January 2009): 95–136.

53. The site is called Venta del Moro. See Jorge Morales, Pablo Paláez-Campomanes, Juan Abella, et al., "The Ventian Mammal Age (Latest Miocene): Present State," *Spanish Journal of Paleontology* 28, no. 2 (2013): 149–160.

54. The modern dromedaries of the Sahara are domesticated camels from Asia. They were first brought to northern Africa by people two thousand years ago.

55. Ferhat Kaya, Faysal Bibi, Indrė Žliobaitė, et al., "The Rise and Fall of the Old World Savannah Fauna and the Origins of the African Savannah Biome," *Nature Ecology and Evolution* 2 (2008): 241–246.

56. Zebras, like horses and donkeys, belong to the genus *Equus*. This genus arose in North America and arrived in Africa by way of Asia.

57. There were still lions in Europe until historical times, as evidenced by sagas such as the Nibelungenlied, coats of arms, sculptures, and paintings. Lions were not extirpated from the Middle East until about two hundred years ago. Hyenas, too, survived and continue to survive in small remnant populations in areas around the Caspian Sea.

58. This does not apply to the same extent to the plants of the African savannah, which developed in Africa.

Part 5: What Makes Humans Human

1. Dierk Suhr, *Mosaik der Menschenwerdung* (Berlin: Springer, 2018).

2. Rüdiger Braun, *Unsere 7 Sinne—Die Schlüssel zur Psyche* (Munich: Kösel, 2019).

3. We do not have a complete thumb from Udo; however, the central thumb bone is relatively large when compared with the length of his index and ring fingers.

4. Sergio Almécija, Jeroen B. Smaers, William L. Jungers, et al., "The Evolution of Human and Ape Hand Proportions," *Nature Communications* 6, no. 7717 (July 14, 2015).

5. Louis S. B. Leakey, P. V. Tobias, and J. R. Napier, "A New Species of the Genus *Homo* From Olduvai Gorge," *Nature* 202 (1964): 7-9.

6. John Napier, *Hands* (Princeton: Princeton University Press, 1993), 55.

7. Kathy Schick and Nicholas Patrick Toth, *Making Silent Bones Speak: Human Evolution and the Dawn of Technology* (New York: Simon & Schuster, 1993), 168.

8. Shannon P. McPherron, Zeresenay Alemseged, Curtis W. Marean, et al., "Evidence for Stone-Tool Assisted Consumption of Animal Tissues Before 3.39 Million Years Ago at Dikika, Ethiopia," *Nature* 466 (2010): 857-860.

9. Sonia Harmand, Jason E. Lewis, Craig S. Feibel, et al., "3.3 Million-Year-Old Stone Tools From Lomekwi 3, West Turkana, Kenya," *Nature* 521 (2015): 310-315.

10. Matthew H. Skinner, "Human-Like Hand Use in *Australopithecus africanus*," *Science* 347 (2015): 395-399.

11. Quoted in Max Planck Institute, "Early Human Ancestors Used Their Hands Like Modern Humans," news release (January 22, 2015).

12. David McNeill, *How Language Began: Gesture and Speech in Human Evolution* (Cambridge: Cambridge University Press, 2012).

13. Kirsty E. Graham, "Bonobo and Chimpanzee Gestures Overlap Extensively in Meaning," *PLOS Biology* 16, no. 2 (February 27, 2018).

14. Michael Tomasello, *The Origins of Human Communication* (Cambridge, MA: MIT Press, 2008).

15. Interview with Robert Brammer on the radio program "Im Anfang war die Geste—Vom Ursprung der Sprache," SWR2 Wissen (April 19, 2010).

16. David McNeill, *Gesture and Thought* (Chicago: University of Chicago Press, 2005).

17. Rüdiger Braun, *Der Menschenplanet* (Frankfurt: Gesellschaft Deutscher Chemiker, 2015).

18. Pesse is a small village in the Dutch province of Drenthe. The canoe was found in 1955. It is 10 feet (3 meters) long and was hacked out of the trunk of a pine tree using stone tools. The boat is embedded in the peat layers of what were once fenlands. See Michiel Gerding, *Het Drenthe boek* (Waanders, Zwolle: Drents Archief, 2007).

19. Peter Brown, Thomas Sutikna, Michael J. Morwood, et al., "A New Small-Bodied Hominin From the Late Pleistocene of Flores, Indonesia," *Nature* 431 (2004): 1055–1061.

20. Fachroel Aziz, Michael J. Morwood, and Gerrot D. van den Bergh, eds., *Pleistocene Geology, Palaeontology, and Archaeology of the Soa Basin, Central Flores, Indonesia* (Bandang: Geological Survey Institute, 2008).

21. Daisuke Kubo, Reiko T. Kono, and Yousuke Kaifu, "Brain Size of *Homo floresiensis* and Its Evolutionary Implications," *Proceedings of the Royal Society B* 280 (June 7, 2013).

22. William L. Jungers, Susan G. Larson, William E. H. Harcourt-Smith, et al., "Descriptions of the Lower Limb Skeleton of *Homo floresiensis*," *Journal of Human Evolution* 57 (2009): 538–554.

23. Matthew W. Tocheri, "The Primitive Wrist of *Homo floresiensis* and Its Implications for Hominin Evolution," *Science* 317 (2007): 1743–1745.

24. Susan G. Larson, William L. Jungers, Michael J. Morwood, et al., "*Homo floresiensis* and the Evolution of the Hominin Shoulder," *Journal of Human Evolution* 53, no. 6 (2007): 718–731.

25. Thomas Sutikna, Matthew W. Tocheri, Michael J. Morwood, et al., "Revised Stratigraphy and Chronology for *Homo floresiensis* at Liang Bua in Indonesia," *Nature* 532 (2016): 366–369.

26. Thomas Sutikna, Matthew W. Tocheri, J. Tyler Faith, et al., "The Spatio-Temporal Distribution of Archaeological and Faunal Finds at Liang Bua

(Flores, Indonesia) in Light of the Revised Chronology for *Homo floresiensis*," *Journal of Human Evolution* 124 (2018): 52–74.

27. Gerrit D. van den Bergh, Yousuke Kaifu, Iwan Kurniawan, et al., "*Homo floresiensis*-Like Fossils From the Early Middle Pleistocene of Flores," *Nature* 534 (2016): 245–248.

28. Adam Brumm, Gitte M. Jensen, Gerrit D. van den Bergh, et al., "Hominins on Flores, Indonesia, by One Million Years Ago," *Nature* 464 (2010): 748–752.

29. Hanneke J. M. Meijer, Thomas Sutikna, E. Wahyu Saptomo, et al., "Late Pleistocene-Holocene Non-Passerine Avifauna of Liang Bua (Flores, Indonesia)," *Journal of Vertebrate Paleontology* 33 (2013): 877–894.

30. Elisa Locatelli, Rokhus Awe Due, Gerrit D. van den Bergh, and Lars W. van den Hoek Ostende, "Pleistocene Survivors and Holocene Extinctions: The Giant Rats From Liang Bua (Flores, Indonesia)," *Quaternary International* 281 (2012): 47–57.

31. Modern humans first brought wild pigs, hedgehogs, and civet cats to Indonesia a few thousand years ago, followed by house rats, dogs, macaques, deer, cattle, and water buffalo. These animals have since naturalized.

32. The Galápagos Islands are a famous example. There, more than 600 miles (1,000 kilometers) from the coast of South America, you find land-dwelling giant tortoises. The same holds true for the Seychelles, which are similarly far from the mainland. Large mainland lizards have also settled on the Canary Islands. And there is a hippopotamus that lives in Madagascar.

33. Settling new islands is not as difficult for reptiles because they are capable of parthenogenesis, a form of sexual reproduction where you basically need just a single female to found a new population.

34. J. de Vos and J. W. F. Reumer, eds., "Elephants Have a Snorkel!" *Deinsea*, Rotterdam 7 (1999): 3–19.

35. Elephants once lived on Sicily, Crete, Cyprus, and many small islands in the Mediterranean. Like the stegodons on Flores, they all became smaller.

36. Phillip V. Tobias, "An Afro-European and Euro-African Human Pathway Through Sardinia, With Notes on Humanity's World-Wide Water Traversals and Proboscidean Comparisons," *Human Evolution* 17, no. 3 (2002): 157–173.

37. Karen McComb, David Reby, Lucy Baker, et al., "Long-Distance Communication of Acoustic Cues to Social Identity in African Elephants," *Animal Behavior*, no. 3 (2003): 327–329.

38. During multiple global lows in sea level during the ice age, Java, Sumatra, Borneo, and Malaysia were joined to the Asian mainland. The large Southeast Asian peninsula that was formed was called Sundaland.

39. Debbie Argue, Colin P. Groves, Michael S. Y. Lee, and William L. Jungers, "The Affinities of *Homo floresiensis* Based on Phylogenetic Analyses of Cranial, Dental, and Postcranial Characters," *Journal of Human Evolution* 107 (June 2017): 107–133.

40. Manuel Will, Adrián Pablos, and Jay T. Stock, "Long-Term Patterns of Body Mass and Stature Evolution Within the Hominin Lineage," *Royal Society Open Science* 4 (November 8, 2017).

41. Armand Salvador Mijares, Florent Détroit, Philip Piper, et al., "New Evidence for a 67,000-Year-Old Human Presence at Callao Cave, Luzon, Philippines," *Journal of Human Evolution* 59 (2010) 123–132.

42. Florent Détroit, Armand Salvador Mijares, Julien Corny, et al., "A New Species of *Homo* From the Late Pleistocene of the Philippines," *Nature* 360 (2019): 181–186.

43. T. Ingicco, Gerrit D. van den Bergh, C. Jago-on, et al., "Earliest Known Hominin Activity in the Philippines by 709 Thousand Years Ago," *Nature* 557 (2018): 233–237.

44. Lizzie Wade, "New Species of Ancient Human Unearthed in the Philippines," *Science* (April 10, 2019).

45. Matthew W. Tocheri, "Previously Unknown Human Species Found in Asia Raises Questions About Early Hominin Dispersals From Africa," *Nature* 568 (2019): 176–178.

46. There is further evidence of earlier, perhaps even much earlier, ocean journeys. Stone tools that could have been manufactured in early Paleolithic times have been found on Crete and on Socotra, a small island in the Gulf of Aden between Somalia and Yemen. The finds from Socotra, in particular, could yield particularly interesting information. The island lies halfway between the Horn of Africa and the Arabian Peninsula and the finds are similar to tools of the Oldowan culture in East Africa that have been linked to an early member of the genus *Homo*. However, a full scientific analysis of the collections is still a long way off, and the artifacts could also have been formed by natural processes. See Curtis Runnels, Chad DiGregorio, Karl Wegmann, et al., "Lower Paleolithic Artifacts From Plakias, Crete: Implications of Hominin Dispersals," *Eurasian Prehistory* 11 (2015): 129–152, and S. V. Aleksandrovic, "Researching the Stone Age on Socotra" academic report in Russian (2010).

47. This is the Self-Transcendence 3100 Mile Race in the New York borough of Queens. It is held on a circuit around an extended city block.

48. Christopher McDougall, *Born to Run: A Hidden Tribe, Superathletes, and the Greatest Race the World Has Never Seen* (New York: Knopf, 2009).

49. Pat Shipman, "How Do You Kill a Mammoth? Taphonomic Investigations of Mammoth Megasites," *Quaternary International* 359/360 (2015): 38–46.

50. Dennis M. Bramble and Daniel E. Lieberman, "Endurance Running and the Evolution of *Homo*," *Nature* 432 (2004): 345–352.

51. Campbell Rolian, Daniel E. Lieberman, Joseph Hamill, et al., "Walking, Running, and the Evolution of Short Toes in Humans," *Journal of Experimental Biology* 212 (2008): 713–721.

52. David A. Raichlen, Hunter Armstrong, and Daniel E. Lieberman, "Calcaneus Length Determines Running Economy: Implications for Endurance Running Performance in Modern Humans and Neanderthals," *Journal of Human Evolution* 60 (2011): 299–308.

53. Rüdiger Braun, *Unsere 7 Sinne—Die Schlüssel zur Psyche* (Munich: Kösel, 2019).

54. Charles Darwin, *The Descent of Man, and Selection in Relation to Sex* (London: John Murray, 1871).

55. Finds of rocks containing the iron-bearing mineral pyrite that could be struck against flint to make sparks are evidence of early firelighters. This method has been used by modern humans for at least thirty-two thousand years. Current investigations by Dutch and French scientists suggest that Neanderthals mastered the art of making fire fifty thousand years ago. It is possible that Neanderthals taught modern humans how to make fire.

56. Andrew C. Sorensen, Emile Claud, and M. Soressi, "Neanderthal Fire-Making Technology Inferred From Microwear Analysis," *Scientific Reports* 8, no. 10065 (July 19, 2018).

57. Francesco Berna, Paul Goldberg, Liora Kolska Horwitz, et al., "Microstratigraphic Evidence of *In Situ* Fire in the Acheulean Strata of Wonderwerk Cave, Northern Cape Province, South Africa," *Proceedings of the National Academy of Sciences of the United States of America* (May 15, 2012).

58. Richard Wrangham, *Catching Fire: How Cooking Made Us Human* (New York: Basic Books, 2009).

59. Jerry Adler, "Why Fire Makes Us Human," *Smithsonian Magazine* (June 2013).

60. Richard Wrangham, *Catching Fire: How Cooking Made Us Human* (New York: Basic Books, 2009).

61. Richard Wrangham, *Catching Fire: How Cooking Made Us Human* (New York: Basic Books, 2009), 98.

62. Jochen Fuss, Gregor Uhlig, and Madelaine Böhme, "Earliest Evidence of Caries Lesion in Hominids Reveal Sugar-Rich Diet for a Middle Miocene Dryopithecene From Europe," *PLOS ONE* 13, no. 8 (August 30, 2018): e0203307.

63. Karen Hardy, Jennie Brand-Miller, Katherine D. Brown, et al., "The Importance of Dietary Carbohydrate in Human Evolution," *Quarterly Review of Biology* 90, no. 3 (September 2015): 251–268.

64. Karina Fonseca-Azevedo and Suzana Herculano-Housel, "Metabolic Constraint Imposes Tradeoff Between Body Size and Number of Brain Neurons in Human Evolution," *PNAS* 109, no. 45 (November 6, 2012): 18571–18576.

65. Dieter E. Zimmer, *So kommt der Mensch zur Sprache* (Munich: Heyne, 2008).

66. Dorothy Cheney and Robert Seyfarth, *How Monkeys See the World* (Chicago: University of Chicago Press, 1990).

67. Steven Pinker, *The Language Instinct* (New York: Morrow, 1994).

68. Chip Walter, *Thumbs, Toes, and Tears* (New York: Walker, 2006).

69. Johan J. Bolhuis, Ian Tattersall, Noam Chomsky, and Robert C. Berwick, "How Could Language Have Evolved?" *PLOS Biology* 12, no. 8 (August 26, 2014): e1001934.

70. Richard F. Green, Johannes Krause, Adrian W. Biggs, et al., "A Draft Sequence and Preliminary Analysis of the Neandertal Genome," *Science* 328, no. 5979 (May 7, 2010): 710–722.

71. The Forkhead-Box-P2 gene, FOXP2 for short, was first discovered in the 1990s in members of a London, UK, family who all suffered from severe speech disorders. Even small changes in this gene have significant effects, because it regulates the interactions of numerous other genes. FOXP2 encodes a transcription factor, a protein complex, that regulates the on and off switches of other genes by adhering to specific areas of genetic material.

72. Weiguo Shu, Julie Y. Cho, Yuhui Jiang, et al., "Altered Ultrasonic Vocalization in Mice With a Disruption in the *Foxp2* Gene," *PNAS* 102, no. 27 (July 5, 2005): 9643–9648.

73. Eriko Fujita, Yuko Tanabe, Akira Shiota, et al., "Ultrasonic Vocalization Impairment of Foxp2 (R552H) Knockin Mice Related to Speech-Language

Disorder and Abnormality of Purkinje Cells," *PNAS* 105, no. 8 (February 26, 2008): 3117–3112.

74. Dan Dediu and Stephen C. Levinson, "Neanderthal Language Revisited: Not Only Us," *Current Opinion in Behavioral Sciences* 21 (June 2018): 49–55.

75. Dan Dediu and Stephen C. Levinson, "On the Antiquity of Language: The Reinterpretation of Neandertal Linguistic Capacities and Its Consequences," *Frontiers of Psychology* (July 5, 2013).

76. Robin Dunbar, *Grooming, Gossip, and the Evolution of Language* (Cambridge: Harvard University Press, 1996).

77. Michael Tomasello, *Why We Cooperate* (Cambridge, MA: MIT Press, 2009).

Part 6: The Lone Survivor

1. Mark Grabowski, Kevin Hatala, William Jungers, and Brian Richmond, "Body Mass Estimates of Hominin Fossils and the Evolution of Human Body Size," *Journal of Human Evolution* 85 (August 2015): 75–93.

2. Ann Gibbons, "A New Body of Evidence Fleshes Out *Homo erectus*," *Science* 317, no. 5845 (September 2007): 1664.

3. Robin Dennell and Wil Roebroeks, "An Asian Perspective on Early Human Dispersal From Africa," *Nature* 438 (December 22, 2005): 1099–1104.

4. David Lordkipanidze, Abesalom Vekua, Reid Ferring, et al., "The Earliest Toothless Hominin Skull," *Nature* 434 (May 1, 2005): 717–718.

5. Originally proposed as a separate species, *Homo ergaster* is now mostly considered an African variety of *Homo erectus*.

6. From 2.6 million to 0.7 million years ago, the cycle lasted for about 40,000 years. After that, a rhythm of 90,000 years of glaciation followed by 10,000 years of warming set in and continues to this day. Science does not yet have a full explanation for what drives these fluctuations.

7. Kim A. Jakob, "Late Pliocene to Early Pleistocene Millennial-Scale Climate Fluctuations and Sea-Level Sustainability: A View From the Tropical Pacific and the North Atlantic," thesis, University of Heidelberg (2017): 212.

8. Xin Wang, Haitao Wei, Mehdi Taheri, et al., "Early Pleistocene Climate in Western Arid Central Asia Inferred From Loess-Palaeosol Sequences," *Scientific Reports* 6, no. 20560 (February 3, 2016).

9. R.-D. Kahlke, "The Origin of the Eurasian Mammoth Faunas (*Mammuthus-Coelodonta* Faunal Complex)," *Quaternary Science Reviews* 96 (July 15, 2014): 32–49.

10. Bienvenido Martínez-Navarro, "Early Pleistocene Faunas of Eurasia and Hominin Dispersals," in John G. Fleagle, John J. Shea, Frederick E. Grine, et al., eds., *Out of Africa I: The First Hominin Colonization of Eurasia (Vertebrate Paleobiology and Paleoanthropology)* (Springer, 2010): 207–224.

11. Robin Dennell and Wil Roebroeks, "An Asian Perspective on Early Human Dispersal From Africa," *Nature* 438 (December 22, 2005): 1099–1104.

12. New research shows that Denisova Cave was visited by Denisovans at least 200,000 years ago and that they survived in this area until about 50,000 years ago. Neanderthals mostly visited the cave between 200,000 and 100,000 years ago. See Katerina Douka, Viviane Slon, Zenobia Jacobs, et al., "Age Estimates for Hominin Fossils and the Onset of the Upper Palaeolithic at Denisova Cave," *Nature* 565 (January 30, 2019): 640–644.

13. Lu Chen, Aaron B. Wolf, Wenqing Fu, et al., "Identifying and Interpreting Apparent Neanderthal Ancestry in African Individuals," *Cell* 180, no. 4 (February 2020): 677–687.e16.

14. In 2014, papers were published that showed some peoples on mainland Asia and even a few groups in South America carry about 0.2 percent Denisovan DNA. But that is a vanishingly small amount compared to the much more significant traces in Papua, a few Southeast Asian islands, and Australia. See Kay Prüfer, Fernando Racimo, Nick Patterson, et al., "The Complete Genome Sequence of a Neanderthal From the Altai Mountains," *Nature* 505 (January 2014): 43–49.

15. The finds of *Homo floresiensis* on the Indonesian island of Flores, also known as the Hobbit, and of *Homo luzonensis* on the Philippine island of Luzon showed that early *Homo*, or even perhaps early hominins, might have been seafarers.

16. The Wallace Line was named after the British biologist Alfred Russel Wallace, who explored this region from 1854 to 1862.

17. Guy S. Jacobs, Georgi Hudjashov, Lauri Saag, et al., "Multiple Deeply Divergent Denisovan Ancestries in Papuans," *Cell* 177, no. 4 (May 2, 2019): 1010–1021.

18. Robin McKie, "Meet Denny, the Ancient Mixed-Heritage Mystery Girl," *Guardian*, November 24, 2018.

19. Viviane Slon, Fabrizio Mafessoni, Benjamin Vernot, et al., "The Genome of the Offspring of a Neanderthal Mother and a Denisovan Father," *Nature* 561 (August 22, 2018): 113–116.

20. Emilia Huerta-Sánchez, Xin Jin, Asan, et al., "Altitude Adaptation in Tibetans Caused by Introgression of Denisovan-Like DNA," *Nature* 512 (July 2, 2014): 194–197.

21. Fernando Racimo, David Gokhman, Matteo Fumagalli, et al., "Archaic Adaptive Introgression in *TBX15/WARS2*," *Molecular Biology and Evolution* 34, no. 3 (March 2017): 509–524.

22. The disappearance of the megafauna in Madagascar—there was an elephant bird that weighed more than 1,750 pounds (800 kilograms) and a 450-pound (200-kilogram) giant lemur—also coincided with the arrival of humans on that island over 1,500 years ago.

23. Paul Martin Schultz, "Prehistoric Overkill," in Paul S. Martin and H. E. Wright, eds., *Pleistocene Extinctions* (New Haven: Yale, 1967).

24. Here, however, nature presents researchers with an enormous challenge. Molecular traces are preserved in fossil material only in cold, dry conditions. We will probably never be able to extract fossil DNA from tropical regions and so any picture we do manage to reconstruct will, of necessity, be incomplete.

25. These genetic, morphological, and cultural entities used to be collectively referred to as human races. The term "race" has been discredited after theories about race were misused for political ends in the twentieth century. See Hynek Burda, Peter Bayer, and Jan Zrzavý, *Humanbiologie* (Stuttgart: Ulmer, 2014).

26. In Europeans in the ice age, the amount was 6 percent. See Qiaomei Fu, Cosimo Posth, Mateja Hajdinjak, et al., "The Genetic History of Ice Age Europe," *Nature* 534 (June 9, 2016): 200–205.

27. Joseph Lachance, Benjamin Vernot, Clara C. Elbers, et al., "Evolutionary History and Adaptation From High-Coverage Whole-Genome Sequences of Diverse African Hunter-Gatherers," *Cell* 150 (2012): 457–469. And, Michael F. Hammer, August E. Woerner, Fernando L. Mendez, et al., "Genetic Evidence for Archaic Admixture in Africa," *Proceedings of the National Academy of Sciences* 108, no. 37 (September 3, 2011): 5123–15128.

28. Cosimo Posth, Christoph Wißing, Keiko Kitagawa, et al., "Deeply Divergent Archaic Mitochondrial Genome Provides Lower Time Boundary for African Gene Flow Into Neanderthal," *Nature Communications* 8, no. 16046 (July 4, 2017).

29. Michael D. Gregory, J. Shane Kippenhan, Daniel P. Eisenberg, et al., "Neanderthal-Derived Genetic Variation Shapes Modern Human Cranium and Brain," *Nature Scientific Reports* 7 (July 24, 2017).

30. Philipp Gunz, "Neandertal Introgression Sheds Light on Modern Human Endocranial Globularity," *Current Biology* 29, no. 1 (January 7, 2019): 120–127.

31. Evgeny E. Akkuratov, Mikhail S. Gelfand, and Ekaterina E. Khrameeva, "Neanderthal and Denisovan Ancestry in Papuans: A Functional Study," *Journal of Bioinformatics and Computational Biology* 16, no. 2 (2018): 1840011.

32. Rebecca R. Ackermann, Alex Mackay, and Michael L. Arnold, "The Hybrid Origin of 'Modern' Humans," *Evolutionary Biology* 43, no. 1 (March 2015): 1–11.

33. Colin Barras, "Who Are You? How the Story of Human Origins Is Being Rewritten," *New Scientist* (August 23, 2017).

34. Kay Prüfer, Cesare de Filippo, Steffi Grote, et al., "A High-Coverage Neandertal Genome from Vindija Cave in Croatia," *Science* 358, no. 6363 (November 3, 2017): 655–658.

35. Richard E. Green, Anna-Sapfo Malaspinas, Johannes Krause, et al., "A Complete Neanderthal Mitochondrial Genome Sequence Determined by High-Throughput Sequencing," *Cell* 134, no. 3 (August 8, 2008): 416–426.

36. Jean-Jacques Hublin and Wil Roebroeks, "Ebb and Flow or Regional Extinctions? On the Character of Neandertal Occupation of Northern Environments," *Comptes Rendus Palevol* 8, no. 5 (2009): 503–509.

37. Robin Dennell, María Martinón-Torres, José Castro, et al., "Hominin Variability, Climatic Instability and Population Demography in Middle Pleistocene Europe," *Quaternary Science Reviews* 30, nos. 11–12 (June 2011): 1511–1524.

ILLUSTRATION
CREDITS

Photographs

14 Bavarian State Collection for Paleontology and Geology (Munich)

17 Madelaine Böhme

122 Alain Beauvilain

123 Agnes Fatz/University of Tübingen

168 Madelaine Böhme

214 Madelaine Böhme

267 Cantonal Museum of Geology, MGL 95212. Wikimedia © Rama, Creative Commons Attribution-ShareAlike 3.0 France.

Diagrams

All diagrams are © Nadine Gibler. Additional source material is given below.

32 Jochen Fuss, et al., "Potential Hominin Affinities of *Graecopithecus* From the Late Miocene of Europe," *PLOS ONE* 12, no. 5 (May 22, 2017): e0177127.

34 "Formation of Magnetic Anomalies at a Mid-Ocean Ridge," https://earthref. org/ERDA/212/.

64–65 Richard R. Scotese, PALEO Map Project, http://www.scotese.com/.

103 Daniel L. Gebo, "Plantigrady and Foot Adaptation in African Apes: Implications for Hominid Origins," *American Journal of Physical Anthropology* 89, no. 1 (September 1992): 29–58.

104 Gerard D. Gierliński, et al., "Possible Hominin Footprints From the Late Miocene (c. 5.7 Ma) of Crete?" *Proceedings of the Geologists' Association* 128, nos. 5–6 (October 2017): 697–710.

125 Milford Wolpoff, et al., "An Ape or the Ape: Is the Toumaï Cranium TM 266 a Hominid?" *PaleoAnthropology* (January 2006): 36–50.

172 Madelaine Böhme, et al., "Messinian Age and Savannah Environment of the Possible Hominin *Graecopithecus* From Europe," *PLOS ONE* 12, no. 5 (May 22, 2017): e0177347.

186–187 Wout Krijgsman, et al., "The Gibraltar Corridor: Watergate of the Messinian Salinity Crisis," *Marine Geology* 403 (2018): 238–246.

Color Section

1 Velizar Simeonovski

2 (top) Wolfgang Gerber/University of Tübingen

2 (bottom) Jochen Fuss/University of Tübingen with inset by Wolfgang Gerber/University of Tübingen

3 (top and bottom) Agnes Fatz/University of Tübingen

4 (top and bottom) © Nadine Gibler. Adapted from Gerard D. Gierliński, et al., "Possible Hominin Footprints From the Late Miocene (c. 5.7 Ma) of Crete?" *Proceedings of the Geologists' Association* 128, nos. 5–6 (October 2017): 697–710.

5 (top left) Florian Breier (top right) Madelaine Böhme

5 (bottom) © Nadine Gibler. Adapted from C. Owen Lovejoy, et al., "Combining Prehension and Propulsion: The Foot of *Ardipithecus ramidus*," *Science* 326, no. 5949 (October 7, 2009): 72.

6 and 7 © Nadine Gibler. Including photographic material provided by Madelaine Böhme, Agnes Fatz/University of Tübingen, Lanmas/Alamy Stock Photo, and iStock/Creativemarc. Adapted from a graphic in "Woher kommt der Mensch? Forscher zeichnen ein neues Bild von unseren Anfängen" ('Where did humanity come from? Researchers draw a new picture of our beginnings'), *Stern*, December 14, 2017, pages 52–53.

8 Velizar Simeonovski

INDEX

Figures and photographs indicated by page numbers in italics

feet, 105; height, 223; *Homo georgicus* and, 268; *Homo habilis* and, 297n13; Java Man, 46–47, 145, 222; language and, 252–53; Out of Africa theory and, 145; Peking Man, 51; running and, 231

Homo ergaster, 145, 240, 297n15, 305n31, 318n5

Homo floresiensis (Hobbit): introduction, 210; additional finds in So'a Basin, 216; anatomical features, 213–14, *214*; animals found with, 217–19; coexistence with other species, 281; dating, 215–16; discovery and unearthing of, 213, 216–17; disease theory, 215; in family tree, 265; Flores Island and Liang Bua Cave context, 210–13; language and, 255; ocean travel, 220–21, 319n15; origin of, 221–23; Out of Africa theory and, 144–45

Homo georgicus, 237, 266–68, *267*

Homo habilis (Handy Man), 53, 203–4, 222, 265, 297n11, 297n13, 305n31, 306n32

Homo heidelbergensis (Heidelberg Man), 47–48, 255, 281, 296n6

Homo luzonensis, 144, 224–26, 265, 281, 319n15

Homo naledi, 281

Homo neanderthalensis. See Neanderthals

homoplasy, 98–99

Homo rudolfensis, 265, 305n31, 306n32

Homo sapiens (humans): introduction, 281–82; agriculture, 287; coexistence with other species, 281, 285; extinction caused

by, 282–83, 285–86, 320n22; in family tree, 58; on Flores Island, 216; future challenges, 287–88; genes inherited from Denisovans, 275–76, 278–79, 284, 319n14; genes inherited from Neanderthals, 275, 284, 320n26; genes inherited from other species, 283–85; genes shared with other primates, 59; Out of Africa theory and, 55; running and, 231, 234. *See also* diet; hands; language; running; tools

Homo wushanensis (Wushan Man), 143–44, 223, 265

hook grip, 202

Hoploaceratherium belvederense (rhinoceros), 86, 300n44

horses: *Equus* genus, 270; forest horses, 87, 301n54; *Hipparion* genus, 165

Hublin, Jean-Jacques, 206

human evolution: approach to, 3–4, 291–92; author's background, 1–2, 7–8; in Europe, 56, 76; evolutionary position, challenges determining, 98–100, 146–47; family tree, 57–58, *58*, *133*; geological epochs and, *77*; and global climate and ecosystems, *138–39*; ice age and, 93, 140, 269–70; multiregional origin model, 146; Out of Africa theory, 139–41, 142–45, 145–46, 226–27, 266; overview, 59, 260; recent discoveries, 2–3, 289–90; savannah hypothesis, 134–37; Savannahstan theory, 271, 291–92; split from chimpanzees, 55, 58, 59, 115, 116, 137, 296n10; split from

Lemuria, 45-46
lesser apes (gibbons), 58, 299n29,
302n65, 303n2
Liang Bua Cave (Flores Island), 211,
212-13, 215-16, 217, 222. See
also *Homo floresiensis* (Hobbit)
Libby, Willard Frank, 307n3
Lieberman, Daniel, 229
Lindenberg, Udo, 81, 83
Lindermayer, Anton, 13, 15
lions, 153, 194, 312n57
Longgupo Cave (China), 142-44
Lucy (*Australopithecus afarensis*), 54,
90, 105-6, 114, 132, 223, 233,
297n14
Luzon (Philippines), 224-26. See
also *Homo luzonensis*

Macchiarelli, Roberto, 126, 127, 128
Madagascar, 314n32, 320n22
magnetostratigraphy, 33-35, *34*, 155,
299n39
mammoths, 270
marabou storks, 217, 219
Markov, Georgi, 9-10
Martin, Paul Schultz, 282
Max Planck Institute for Evolutionary Anthropology, 205-6, 207,
273, 274-75
Max Planck Institute for Psycholinguistics, 254-55
McNeill, David, 208
meat, 204, 238-39, 243
Mediterranean Sea. *See* Messinian
Salinity Crisis
Mediterranean shrub savannah,
171-73
megafauna, 282-83, 320n22
Mesopithecus pentelicus, 13, *14*, 16,
193
Messinian Salinity Crisis: introduction, 185, 310n42; cause

of, *186-87*, 188; discovery,
186; end of, 192; evaporation
and salt deposits, 188-90,
311nn47-48; geological consequences, 190-92; migration
triggered by, 192-94; significance of, 290-91; Trachilos
footprints and, 111-12
metabolism, 71-72, 236-37
microfossils, 156-57
Microstonyx (pig genus), 164
migration: between Africa and
Eurasia, 62-63, 66, 68, 183,
192-94, 271, 291; assumptions
about, 209-10; from climate
change, 181-82, 183-84, 290-
91; Darwin on, 45; factors for
successful migration, 219-20;
from Messinian Salinity Crisis,
192-94; Out of Africa theory,
139-41; Out of Africa theory,
challenges to, 142-46, 226-27,
266; seafaring, 220-21, 315n46,
319n15; wanderlust, 227
Millennial Man. See *Orrorin*
tugenensis
Miocene, 45, 296n4
Miocene Climatic Optimum, 60,
63, 67, 298n18
Miophasaneus genus (pheasants),
86, 300n45
Miotragocerus monacensis
(Munich wood antelope), 86,
300nn40-41
Mission Paléontologique Franco-
Tchadienne, 119, 129, 304n9
mitochondrial DNA, 274
molecular clock, 114-15, 297n16
monkeys: New World, 57; Old
World, 57, 61
Mount Pentelicon (Greece), 13, 15.
See also Pikermi (Greece)